Ernst Aebi

SIMON & SCHUSTER

New York London

Toronto Sydney

Tokyo Singapore

Seasons
of
$S$and

SIMON & SCHUSTER
Rockefeller Center
1230 Avenue of the Americas
New York, New York 10020

Copyright © 1993 by Ernst Aebi

Designed by Bonni Leon-Berman
Manufactured in the United States of America

10 9 8 7 6 5 4 3 2 1

Library of Congress Cataloging-in-Publication Data

Aebi, Ernst.
Seasons of sand / Ernst Aebi.
p.   cm.
   1. Economic development projects—Mali—Araouane. 2. Agricultural development projects—Mali—Araouane. 3. Agricultural
assistance—Mali—Araouane. 4. Aebi, Ernst—Journeys—Mali.
5. Araouane (Mali)—Description and travel. I. Title.
HC1035.Z7A732   1993
338.96623—dc20   93-2352
                                                                CIP

ISBN   0-671-76935-9

Acknowledgments

All the names of persons appearing in the book are the real ones. I thank those who deserve thanks and occasionally consider forgiving those who didn't.

Some people who don't figure in the story deserve mention. Jonah Blank, author of the Indian travel epic *Arrow of the Blue-Skinned God,* managed to make my manuscript read as if I knew English well and as if I were a good writer; many, many thanks to him. Suzanne Gluck, my agent at ICM, deserves thanks for realizing that Jonah's mind and mine work in similar mysterious ways and thus getting us together. Rebecca Saletan, my editor at Simon & Schuster, deserves thanks for many excellent suggestions; she is one of the few people from whom I gladly accept criticism because it is always constructive. A special thanks goes to Jeri Biehl, who makes it possible for me to roam about in the world by taking care of all my affairs at home.

And I thank Araouane, for many obvious reasons but most of all for allowing me to get to know Emilie.

CONTENTS

"That's all very well,"

Candide said,

"but now we must go

work in the garden."

—Voltaire, *Candide*

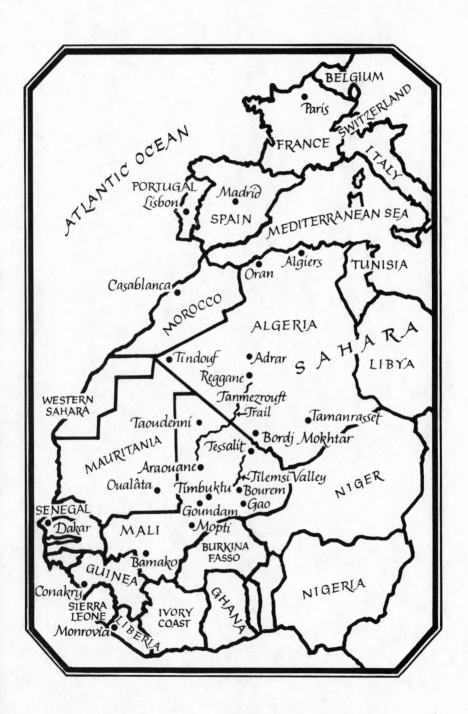

Seasons
of
Sand

P r o l o g u e

It had not rained in Araouane for forty-two years. The town's thirty clay huts were crumbling under the steady sandblasting of the desert winds. Some of them barely managed to keep the encroaching dunes off their roofs. The closest inhabited place was Timbuktu, 260 kilometers to the south. The only vegetation most of the villagers had ever seen was the camel thorn in the courtyard of the mosque. None of them knew how old it was. They could only say that the solitary tree had always been there.

February, 1991

My fiancée, Emilie, and I were playing Scrabble. The game had been going on for hours. Evenings in Araouane were times for relaxation from the hard work of the day. Most of the villagers' talk nowadays was about war: the international conflict in the Persian Gulf; the recent military coup d'état in Bamako that had unseated an unpopular dictator who had ruled Mali for decades; and, even closer at hand, the uprising of the Tuaregs.

Since 1990 the Tuaregs had been fighting a guerrilla campaign against the Malian army. The Tuaregs are a nomadic Berber people numbering only about a million who roam 2 million square miles of the Sahara, and do not like to obey any government. The Arabs called them "the veiled people" because most of their men keep their faces covered. The French called them "the blue people" from the color of their flowing robes known as jellabas. Neither could ever wholly subdue them. For centuries they'd lived in an uneasy balance alongside the Sorai, black Muslims from a tribe that had once ruled the whole region. (The Sorai population of Araouane remember when, a few short generations ago, the Tuaregs used to raid Saharan villages and kidnap whatever blacks they could capture for the slave trade.) The current revolt was sparked when the Malian authorities began siphoning off food shipments that relief agencies had been sending to the Tuaregs.

We were drinking grenadine syrup with water—since the rebels had taken to attacking all vehicles coming through this part of the Sahara, we couldn't get the occasional bottle of beer or soda anymore. All commerce had stopped, and our supplies had dwindled to a mere trickle. We hadn't been able to go to Timbuktu for six weeks. We could buy sick camels from the destitute nomads who came to our wells, but all other staples were running short. In our warehouse only a few sacks of millet remained for a population of about 150.

The fringes of the Tuaregs' traditional territory was periodically patrolled by the Malian army, but there was far too much land for any government to hold down. It was rumored that soldiers from the south had sent the nomads into flight by poisoning their water holes. This winter, we had already eaten fourteen of the Tuaregs' camels. Emilie collected the animals' eyelashes.

"*Salaam aleikum, Salaam aleikum!*"

We wondered who could be coming to our house so long after midnight. The villagers had recently begun keeping to their homes after dark. Everybody feared that the rebels might suddenly show up at our wells. They had never before come to Araouane. Mohammed Ali, with the radio we'd brought him, was giving daily updates on the death toll from the uprising. Whole villages had been wiped out in a flash.

"*Smilla,*" I called in greeting to the people outside the door.

Five nomads filed into the room and stood along the wall, blinking in the weak glow of the solar-powered light bulb. Their heavy burnouses indicated that they were strangers to our region, for none of the local people wore such thick fabric. They had their hands hidden in the folds of their clothes, their turbans tightly wrapped around their faces. They were filthy.

"Peace be with you."

"Peace be with you also."

"May God protect you."

"May God be with you."

"No evil to you."

"None to you either."

"I trust your health is in the hands of God."

"I hope for yours also."

"I trust you are not tired."

"God has provided us with a safe journey."

"God be praised."

"May your sons be in the hands of God."

"May your sons be in the hands of God as well. . ."

The greetings, as usual, went on for a long time. We spoke in classical Fussha Arabic, a language in which neither they nor I was completely comfortable.

I asked them if they would uncover their faces, so I could see who I was talking to. One man in a dark brown cloak slid part of his turban below his chin. He was old and toothless, with a face like a withered prune. I didn't recognize him, but at least I knew he was a Moorish Arab and not a Tuareg. No Tuareg would ever have uncovered his face simply because I asked him to.

"We have to talk to you," he said. "Excuse us for coming so late, but Babaya said you'd probably not be sleeping yet."

"Come and sit with us," I said. As they squeezed behind the table in their bulky robes, one or two of the others uncovered their faces.

"How can we help you?" My initial fears had lessened, but I was still on my guard.

"We come from Bordj-Mokhtar," the man said. "We have three tons of food. We were heading for Timbuktu, but thought you might wish to buy it."

Smugglers. They had just made the twelve-day trip from

the Algerian border, their camels laden with couscous, flour, macaroni, and dates. Normally they would have sold their goods in the city, but fear of the rebels had brought them to us instead.

As we haggled over the price, slurping sweet tea, I couldn't help thinking about how much things had changed. Only three years ago, these traders would never even have thought of stopping at Araouane. If they had stumbled on the town by mistake, they would have met only a few starving people with nothing to offer for such much-needed foodstuffs. These days life throughout this whole stretch of the Sahara had grown precarious, and even Timbuktu teetered on the brink of famine, but Araouane, at least for the moment, seemed secure.

The town had become again what it had been centuries ago—a flourishing community, now even an oasis, in the very center of the world's largest desert.

I'll tell you how this came to pass.

BACK TO THE SALT MINES

February, 1988

I was three days out of Timbuktu—the End of the World, the City that Never Was, the legendary Farthest Place Away. Since the second morning my behind had been covered with blisters, and I could barely sit on my camel's hump. The desert nomads who led the caravan politely pretended not to notice my discomfort. I wanted to get to the salt mines of Taoudenni and back before the intolerably hot summer arrived. But I'd already learned that this is not a part of the world where anything gets done very quickly.

I had come to Africa just for the fun of it. For the first time in many years, I found I had the luxury of doing exactly as I pleased. My work as an artist (and, more lucratively, as a renovator of lofts in New York City's SoHo) had left me financially independent. After a messy divorce I'd spent a decade raising four children by myself. Now the kids were all grown—Nina, Tony, and Jade were in college, and Tania had just completed a record-setting solo circumnavigation of the globe by sailboat. In my twenties I had traveled the world and eventually emigrated from my native Switzerland to America in an attempt to live out my dreams; thirty years later I found them to have gone a bit stale, and it seemed like a good time to hit the road again, in search of new dreams to live out.

So I took up scuba diving and mountain climbing, crossed oceans in sailboats, went on survival treks in the Arctic, drove in the Paris–Dakar off-road rally. It was while racing through the Sahara at breakneck speed that I decided to come back someday at a more leisurely pace. Amid such poverty, I was ashamed of the purposeless waste of money that the race represented, and I vowed to meet the people whose land I blazed through.

I had always been particularly drawn to the desert, to its adventure, its wildness, and its vast solitude. Whenever I saw a book with the word "desert" in its title or a barren landscape on its cover, I'd invariably want to read it. Such accounts of explorations into wastelands—whether of snow or jungle or especially of sand—had been the escape fantasy of my youth. One book in particular drew me to the Sahara: Richard Trench's *Forbidden Sands,* in which the author recounted his unsuccessful attempt in the midseventies to visit the prisoners in the hellish salt mines of Taoudenni. Trench described how he approached the mines from the north with a camel caravan from Tindouf, but the activities of the Polisario separatists in this stretch of desert now made such a route impossible. I virtually drooled over his adventures on the long voyage, and decided to make an expedition of my own from the south.

But in the Malian capital of Bamako I was told that Taoudenni was off-limits "for security reasons"—one of the world's worst prison camps (according to Amnesty International) is located right nearby. After several days I'd gotten a healthy taste of Saharan bureaucracy. I'd talked with scores of gaudily dressed functionaries half hidden by tremendous stacks of paper, both the men and the documents covered with thick layers of dust. It seemed that the more important a bureaucrat was, the higher his tower of unread paper.

Realizing that I was getting nowhere, I chartered a rickety plane with an unscrupulous Lebanese pilot, and hightailed it for Timbuktu. If all went well, I would be two months in the desert with a caravan, so I looked around for some reading material. The only texts for sale in town were a wide assortment of Qur'ans, naturally in Arabic. My only book was Hemingway's *Green Hills of Africa,* but a journalist who shared my plane ride was willing to give me a back issue of *Gentlemen's Quarterly.*

From the moment I landed in the final gateway to the desert, I was surrounded by a black cloud of flies. Not surprising, since the town was full of rotting heaps of garbage and large piles of excrement. The French colonists had left behind some impressive government buildings, but they had been reduced to a cluster of derelict shells in an unimaginable state of disrepair; yet this ruin of collapsed stairs, caved-in roofs, crumbling walls, and piles of uncollected rubbish still served as offices for the current administration.

Officially, modern Timbuktu has about 12,000 inhabitants. But there are reportedly at least another 12,000 refugees from the advancing desert scattered all over town in makeshift straw shelters. How the aid organizations and the government come up with these numbers is a mystery to me, since very few of the residents and almost none of the refugees have any documents of registry.

Most of the adult males in town seemed to do nothing but sit around in their colorful jellabas sipping tea and talking. The more energetic among them sold whatever came up or down the river, or could be smuggled in from Algeria, Mauritania, or Niger. People wearing tribal clothing engaged in whatever profession was traditional to their tribe. Bozo fishermen sold their smelly smoked fish, poor mothers or sisters of workers in the mines cut big bars of salt into smaller marketable chunks, Tamacheq Bellas hammered old car parts into hoes, hatchets, or teakettles, Bambara women in their bright billowing dresses sold watermelons shipped down from Ségou in little boats called pirogues, Sorai farmers sold their unshelled rice and beans, nomads in threadbare robes herded their goats through the streets trying to trade them for sugar and tea, and Tuaregs, known as blacksmiths, offered their crafts to anybody who looked even vaguely like a tourist.

The place was full of noise, movement, and squalor. Several parts of town were home to hordes of beggars, some with limbs, noses, and eyes rotted away by leprosy, cripples shuffling around idly looking for food or for work. Hordes of children, many on crutches, lay, sat, or huddled about sadly, their bellies distended like pumpkins, their eyes dripping with infection.

This, then, was Timbuktu, the fabled city of old and gold, renowned site of palaces and universities, the cradle of West African culture.

The town's Welcome Wagon was the local police station. Everyone passing through Timbuktu was required to check in with the authorities there. I identified the hut by the pack of idlers in mismatched pieces of uniform lounging in front of it. Loud *souka* music burped from a tape deck submerged in the sand.

The station's interior would have been lighted by a single bare bulb hanging from the ceiling, except that the electricity seldom works in the Sahara. Once my eyes had gotten accustomed to the dark, I could make out partially dressed bodies lying on the floor and napping on a makeshift cot. Nobody got up to greet me, but on my left I noticed the hands of detainees stretching through the bars of a cage. I realized with trepidation that I had just encountered Timbuktu's Finest. I didn't want to think what the room's only chair was for—the chair with metal plates attached to electrical wires.

After repeating "*bonjour, messieurs*" until I roused one of the officers from his slumber, I was allowed to fill out the registration papers. Or, rather, I was allowed to look at a master form with pages of questions asking for biographical details—my mother's maiden name, the birth dates of my paternal grandparents and those of both parents—but the answers had to be written out on blank paper because the authorities had run out of extra copies. One line asked the date on which I'd entered the state of "West Africa," a political entity that has not existed since French colonial times.

I paid the fee of 1,000 West African francs (a sum about equal to three U.S. dollars) the policeman demanded for my passport. Out of habit I asked for a receipt—I was in dire

need of toilet paper for my upcoming caravan trip. The clerk
told me that a receipt would take two or three days, at the
least. He then ordered me to give him a pack of cigarettes, a
T-shirt, or a ballpoint pen.

A thousand francs seemed like extortion enough, and I
balked. Fortunately, the officer was too tired to argue, and he
released me into the hands of the Timbuktiens. Literally, into
their hands. Wherever I walked, a horde of kids would stick
their palms in my face and in my pockets. "*Cadeau, il faut
me donner cadeau,*" they would panhandle in their limited
French. Once I gave a few coins to a particularly miserable-
looking boy, and was lucky to make it out of the ensuing
scramble in one piece. I thought to myself, surely there must
be a way of helping these desperately poor people without
beggaring them further.

The destitute children only added to the bad feeling that
Timbuktu gave me. Dodging the sewage from overhead
drains and jumping over slimy puddles of human and animal
excrement along the dusty alleys, I understood how the explorer René Caillié must have felt when he became the first
European to reach the fabled city over a century earlier. Caillié had been disappointed to find only a pathetic collection
of mud huts in the middle of the desert. When he described
his impressions on returning to France, he was branded a liar
and a charlatan. His audience refused to believe that the legendary city could be such a letdown, and they decided that
he'd never actually visited the place. Only many years later,
after other travelers had come back with the same reports,
did he get his recognition.

When I got settled in at the hotel, a little Moorish boy hanging around outside asked me very politely if he could fix me
tea. Since he spoke French fluently and seemed honest
enough, I decided to have him get me to the men who
arranged caravans. I couldn't very well ask the police, and
they were the only people I'd met in town. In the evening
Alouz and I sat in the sand outside the hotel, and he set the

teakettle boiling. He made the traditional three glasses: the first horribly strong and bitter, the second strong and sweet, the third weak and horribly sweet.

"I would like to join a salt caravan to Taoudenni," I said. "Can you help me contact the right people?"

The boy looked at me like I was crazy, and missed the glass into which he was pouring. "You can't go there," he said. "We're not even supposed to talk about it."

"Okay, Alouz, let's not talk about it. But surely you know somebody who has dealings with the caravans?"

"Well, yes, that would be Salah Baba."

"Where can I find him?"

"Right now he is probably on the dune behind the hotel. He usually goes there for evening prayer."

Salah Baba turned out to be a very impressive, regal-looking man. He stood straight as a ramrod, 6 foot 6 inches tall, with a beard half an inch wide framing his chiseled face. He had a nose like an eagle, piercing black eyes, and a Mediterranean complexion. At first I took him for forty, then I figured sixty, then I just couldn't tell. He was easily the best-dressed man I'd seen in Timbuktu, wearing a richly but not gaudily embroidered jellaba and a snow-white turban wrapped loosely around his head. He looked the part of the greatly respected marabout (Islamic religious leader) I'd later discover he was.

Alouz vanished immediately after pointing out the illustrious man, as if afraid to disturb somebody so important. As it turned out, I did not need him to translate. Salah Baba spoke flawless French, as well as Tamacheq, Hassania (the archaic Arabic that the nomads spoke), Arabic, Sorai, and Bambara; I spoke French, English, German, Italian, Spanish, *Schwiizertütsch* (Swiss German), and some Japanese, but none of the local languages, so we conversed in French.

Salah Baba chain-smoked Libertés, cheap Malian cigarettes that often disintegrate to a mess of tobacco and paper before you can even light them.

"So you want to join a salt caravan to Taoudenni?" he said.

"I see no problem there, but it will cost you dearly. You'll have to get provisions for over a month and pay for the camels to carry them and the men to look after you."

I told him that I wouldn't need any special handlers, I was quite capable of taking care of myself, and that surely I wouldn't need more than a single camel to carry my provisions. I just wanted to tag along.

He cut me short, saying that if I wanted to go, I'd have to let him handle everything and do as I was told. The man was obviously accustomed to calling all the shots.

"So," I said, "supposing we come to an agreement, how soon could I leave?"

"It would have to be very soon," he replied. "We are already far into February, and soon there won't be any more caravans leaving. The hot season is almost upon us. I will find a responsible cameleer and start to arrange the provisioning right away."

I certainly liked his no-nonsense approach. But then came the bombshell. He wanted 600,000 francs—the equivalent of $2,000, or the cost of the ridiculously expensive charter flight from Bamako to Timbuktu. I knew that in Africa the first price asked is seldom the final price paid. But my attempts to bargain only caused Salah Baba to turn away.

"There is no time for games," he said. "If you want to go, give me the money now. If not, go home."

I handed him the cash right there.

Not until much later would I learn that the going rate for a full camel charge of salt bars on the same trip was only six dollars.

"I'll also get you a *boubou*," he said, referring to the cameleer's light blue cotton jellaba. "While on the caravan, you'll have to wear that, otherwise you'll spook the camels; they are not accustomed to seeing Western clothes. One of my daughters will make a list for you of some Hassania words which you may need to communicate with the men of the caravan."

Things moved very quickly with Salah Baba. Over the next

few days he bought provisions and had them taken by donkey to a site outside town where they would be loaded on the camels before I even saw them. When he told me about the large number of donkeys and camels that were needed, I timidly asked him what, after all, I was going to eat during the trip.

"Same as the others," he said. Not until the expedition was under way was I able to see what my $2,000 had bought. When I did, I found I owned a half ton of dates, two hundred kilos of baobab powder (a staple of the local diet), a stack of 50-kilogram bags of sugar, several wooden kegs of tea, and a full metric ton of rice stamped "Gift of the People of the United States of America, NOT TO BE SOLD OR EXCHANGED."

I met with Salah Baba on the dune behind the hotel every evening for several days. He was waiting for a cameleer he considered responsible enough, since he didn't want me to get into any trouble on my journey. One evening he showed up with a very little man in an extremely dirty *boubou*.

"This is Dah Ould Lemine," he said. "He will take you to the salt mines and back. I have given him your provisions and paid him for his services. He is a clan chief and a highly respected man in this part of the desert. He is also a famous caravan leader. He speaks only Hassania, so we will make plans for any contingency before you leave."

Except for his aquiline nose, Dah was the exact opposite of Salah Baba. He had a scrawny turkey neck; his face was small and birdlike. On the very few occasions that I saw him without a turban, I noticed that his ears stuck out at right angles to his head. He wore rubber flip-flops with the heels worn almost completely through. His *boubou* looked as if it could stand on its own from all the dirt on it, but he himself seemed clean. The most striking thing about Dah, however, was his ready smile. In morose Timbuktu, I'd almost forgotten what a smiling face looked like.

I think that almost from the start we both sensed we'd like

each other. Dah would concentrate earnestly on my lips as I spoke to him through Salah Baba, as if by paying close enough attention he might be able to understand the alien language. But there wasn't much I had to tell him. I wanted to join his caravan to Taoudenni, I'd eat what the others ate, rise and sleep when they did, perform any tasks they thought I was capable of, and be treated as much like one of them as possible. I doubt he received more than a fraction of the sum I paid Salah Baba for his services.

The day before we were scheduled to depart, Salah Baba told me that foreigners were not allowed to visit the mines because the site was also a "resting place" for political prisoners. But he was sure the Timbuktu police chief could arrange something, for a certain consideration.

The next afternoon we went to the chief's house, one of the more lavish ones in Timbuktu, with a facade decorated in colorful glazed tiles. We sat in the courtyard on a beautiful Algerian carpet, drank tea and nibbled on small pieces of roasted goat.

If our host had been one of the crew at the police station, he had changed completely. He was lively, courteous, and extremely friendly. After the introductory formalities, he asked what he could do for us, although he certainly knew why we'd come. When I told him, he rolled his eyes and moaned in agony over the immensity of the task. There were many potential problems with my plan, he said, and it would require a great deal of organization to assure my safety.

"One never knows," he said, "what can happen in the middle of the desert when it is found out a *toubab* is traveling unprotected."

Over the third glass of tea (weak but very sweet) he handed me a permit to go to Taoudenni—with proper stamps, signatures and all.

"That will be six hundred thousand francs."

I stared in disbelief. "Are you sure it is not six hundred, or maybe six thousand francs?"

"No, no, it's six hundred thousand francs. You cannot imagine how much work it will be for us to assure your safe passage."

It seemed to be the standard fee, so I reluctantly paid again. At that point, I had little choice.

We left that evening, shortly before sundown. The group I was going to join consisted of Dah Ould Lemine, his elder sons, Hadji and Mohammed, and about forty camels. It was obvious that Dah felt very uncomfortable in the city and wanted to waste no time getting back to his home, the desert. Salah Baba and a few of the merchants who had sold me supplies came out to the dunes to wish us a good voyage. All the members of the expedition stood in a circle, joined hands, and prayed silently for a safe passage. As the setting sun painted everything orange, it made a pretty sight. During my time in the Sahara, I was to see this ritual often. It felt like a cushion between the hustle and bustle of provisioning and preparing and the serenity of the waiting desert.

"Remember now," Salah Baba said to me, "these people are very apprehensive about taking you along. They are worried about what will happen to them if you don't come back safe. They are shy, and know little of the outside world. You have to be patient and understanding with them." Yes, I thought, and they with me.

We plodded over dune after dune, always heading north. Dah was in front, looking for passages that were not too steep, leading the first camel by a rope tied around its lower jaw. The two boys ran up and down the long line of camels—the tail of one animal tied to the lower jaw of the next—adjusting a load here, retying a rope there. The men and animals threw long shadows in the moonlight that danced wildly on the undulating sand. Out of nowhere a young boy joined us, leading a goat by a string. An old man with two camels trudged alongside us for a while, then disappeared into the dark.

While Dah and his sons walked, they had me crated on a

saddle that must have been designed by the devil himself. It was mounted on top of, rather than behind, the camel's hump, which made it completely unstable. It had to be lashed on with a mess of ropes, and still the saddle came loose periodically and had to be reattached. I would have preferred walking, but I didn't feel like challenging the decision. When we stopped for the night at around ten o'clock, I was not allowed to help build the campfire or lend a hand in any other evening chore.

By the afternoon of the second day, my ass was one open sore, I had no water to wash any part of my body, and I was already exhausted from the trek. At dinner (a kettle of rice with some pieces of dried camel meat, to judge by the size of the bones) the nomads noticed that I lay on my belly to eat instead of sitting up. They guessed the reason and could not help laughing. Dah prepared a poultice of ground camel dung, ashes, and water, which he indicated I should pack on my sores. Instead I used up my whole supply of Band-Aids and almost all my precious toilet paper for padding.

Even though we could not speak the same language, the nomads managed to make me feel as if I were one of them. It took me quite a while to understand many of the things they did, why they did them, and what they expected of me, but eventually I settled into their routine.

It was a routine without much variation. Each man had a blanket in which he curled up to sleep. He spread it on top of his camel's cargo to make a softer seat while he was riding, and wrapped it around his body while cooking, eating, or talking in the evening. Around 5 A.M. everybody would rise from the sand. The very first act of the day was the morning prayer and then the preparation of tea, the beverage that seems to be the staff of life to the nomads. If an interrogator wanted to force a nomad to talk, he wouldn't have to use physical torture (they're much too tough for that, anyhow); he'd only have to threaten to pour out his morning tea. After tea came breakfast, usually crumbled bits of dried bread soaked in hot water, with a bit of sugar and *boulenga*.

The flat loaves of bread were carried in goatskins and crammed underneath the baggage that the cameleers sat on, so they soon became a pile of crumbs. The dry climate, however, prevented mold. *Boulenga* is a greasy yellow-gray-black-green substance that comes from some plant in the south and smells like a woolen sock that has been on an unwashed foot inside a rubber boot for a month. The nomads carried it in leather pouches, and they loved it.

After breakfast, the men scattered in all directions to round up the camels. In the evening, the camels were let loose to feed on whatever pasture they could find, which is why we never stopped for the night at any specific time, but whenever we happened to come across an area with suitable vegetation. Depending on how far the camels had wandered and how badly the wind had obscured their tracks, the rounding up and reloading could take anywhere from an hour to several hours.

At the beginning of each day's journey the nomads walked, to make sure that all the baggage was well balanced and well secured. After two or three hours, everyone climbed on a camel and rode sitting on top of the luggage. They varied the camels that they mounted, to avoid wearing out any one animal.

Around midday (depending, again, on the availability of pasture) the caravan would stop, unload all the camels to let them search for food, and prepare tea and *crème*. This concoction could hardly be further from haute cuisine, but for some reason all the nomads (even if they didn't speak another word of the language) called it by this French name. The recipe for *crème* is a fistful of baobab powder, three fistfuls of ground millet, and water. The mixture was stirred by hand, then passed around. Each nomad took a swig of the soupy stuff, and when it was gone, they'd scoop out the dregs with their hands.

After about two hours, the procedures of the morning were repeated: make tea, gather the camels, load them, move out. For the afternoon trek, however, the men generally rode the camels right from the start. Evening was the time for the day's big meal, which had some variety. But whether the dinner was dried camel meat with rice, baobab powder with pounded millet, or stuffed goat guts, the primary ingredient was always sand.

These, then, were the habits of the people I was going to live with for the next month or so. I could not understand a word of the muted conversations they held. There was never any yelling, as if they didn't want to disturb the quiet of the desert (what a change from noisy Timbuktu). Salah Baba had told me that the nomads spoke Hassania, but to my ear their high-pitched melodic utterances had nothing to do with the guttural modern Arabic spoken from Algeria to Egypt and Syria.

Before coming to the region, I'd assumed I would be living with the Tuaregs, the former Berber rulers of the Sahara about whom so many romantic stories had been written. But on reaching the Sahara, I learned that for over 400 years the corridor that ran from Timbuktu through Araouane, Taoudenni, and Tindouf to Morocco had been controlled by Moors. In the early sixteenth century, these people of Arab stock had taken the regions of Timbuktu, Gao, and Djénné away from the black Songhai empire, and replaced the Tuaregs as the masters of the open desert and its caravan routes. Dah and the rest of the nomads with whom I traveled were Moors—people in this region generally use the terms "Moor" and "Arab" interchangeably, since the Moors are simply Arabs originally from the Maghreb in today's Morocco. The nomads in that part of Mali are almost all Moors, and I would not meet any large groups of Tuaregs until much later.

On the afternoon of the third day we heard a car. A Land-Rover appeared over a dune, with a police captain, three heavily armed officers, and a Moorish guide. They pointed their gun barrels at my head and invited me into the front seat. I asked the Moor to tell Dah and his sons to wait for me, to say that I would return before long. Then they turned around and drove me all the way back to Timbuktu.

At the station, a new police chief offered me a chair. Fortunately, it was not the one with steel plates and wires. The officer whose house I'd visited and who had written my Taoudenni permit, I learned, was in jail. I was charged with bribing a police captain, and with possible espionage as well. It sounded like I was going to visit Taoudenni after all—in handcuffs. Cutting salt bars sounded like more adventure than I had bargained for, especially if my fellow miners were to be disgraced Malian bureaucrats and administrators. I'd never gotten along particularly well with pencil pushers.

Two lucky breaks and a lie sprung me. First, the phone line out of Timbuktu worked; the same time the following year it would be out for nearly four months. Then the U.S. embassy in Bamako actually answered the call. I told the embassy switchboard operator that I was working for *National Geographic*. It wasn't strictly true, but I had met an editor there once.

The embassy did something—God knows what—but it worked. After ten frustrating days hanging around the hotel (I was the only guest, and my sole entertainment was drinking tea with Alouz in the sand outside), I got my passport back with a strict warning that if I was found anywhere near the salt mines I was going to stay there.

For $500 I hired a Landcruiser which, from the looks of it, should have been retired long ago, and with the same three policemen and the guide I set out to look for Dah and his caravan. The guide assured me they would have waited. But in the area where we'd left them, some nomads told us the caravan had given up hope of my return and moved on the day before. We followed the trail of camel turds, which now led away from Taoudenni in the direction of Mauritania to the northwest.

As the Landcruiser sputtered on the last of its fuel, we found Dah and his camels. I explained that I was forbidden to go near the salt mines, but I wanted to tag along wherever they were going. We left the police crew searching for a camel to fetch diesel from Timbuktu, at least a three-day trip each way. I took great pleasure in abandoning them to the desert.

My sore behind had healed in Timbuktu, and I was all set to show Dah and his boys that I was not a tenderfoot. We covered about 400 kilometers over the next eight days. Most of the time I walked (to spare my butt further abuse) but sometimes my legs would begin to weary, and I'd ride one of the camels. When I did, however, I made sure to sit on a pile of grain-sacks rather than the devil's saddle.

Mounting and dismounting my camel during the trek was always an ordeal. Because of the way the caravan is strung together, you cannot stop one beast without bringing the whole convoy to a halt. And not having a saddle to grab made it even trickier for me to master the standard Saharan mounting technique: First you walk along the shady side of your camel making low murmuring sounds. When the beast lowers its head to listen to what you have to say, you grab its ear and pull down as hard as you can. The camel bellows like an ambulance crossed with a tugboat, but you have to keep pulling. When the animal finally realizes there will be less pain if it keeps its head down, you put your right foot on its neck, and slowly let the ear go. Instantly the head snaps up, catapulting you onto the load of baggage strung across the hairy hump. But if you release the ear too fast you may get hurled clear over the camel's spine into the sand on the other side.

You are now on, but facing the wrong direction—to use the neck as a springboard, you have to mount backward. Turning around would be simple enough, if the camel weren't still mad at having been tricked into letting you on. It jumps about for the first few minutes, and spitefully tries to shake you off its back.

Once I saw a novel method of mounting. An old man who had come out of nowhere and joined us for a few hours got off his camel to stroll and talk with Dah. When they parted, the man went behind his camel, grabbed the tail as if it were a rope, and walked straight up the beast's hind legs. I tried this several times, but never got farther than the shaggy knee.

Getting off is far simpler—you jump. Whenever I dismounted for any reason, the men of the caravan got down to walk beside me. At first I thought they were making sure I'd landed safely, but later I learned they didn't want to miss the hilarious spectacle when I tried to climb on again.

We crossed endless rows of beautifully formed dunes. The light and shade created magnificent patterns on the mountains of sand. Sensuous shapes invited the mind to wander. Was the landscape moving or were we?

Little mountains of sand formed constantly on big mountains of sand, only to be blown away with the next blast of wind. The prevailing winds were northwesterly, and I wondered whether in time the whole Sahara would be blown into the Atlantic Ocean. When I looked into the distance I could not venture to guess how far away a hill might be, since there was no tree, house, animal or any other fixed point of reference. Huge, small, or minuscule, all dimensions melted into one. One foot in front of the other, each footstep obliterated with the next—did I step into the future or the past?

In many areas the sand was soft and powdery, sculpted by the wind. I strode over small dunes but had to struggle up the bigger ones, sliding back half of every step I took. The dunes looked like convoys of ocean liners. Sometimes the whole drift would start to shift when I put my foot down, burying me up to the thighs in hot dust. I soon found I saved a lot of energy if I stayed close to the nomads and the camels, who somehow knew just which paths would offer the least resistance because they knew where the ground was firm. Yet, when I wandered away from them, into the untouched desert, I did not feel like a despoiler: Nature soon obliterated my footsteps.

It reminded me of the sea, where I'd last experienced true solitude while crossing the Atlantic in a small sailboat. Alone on the vast water, I'd felt my place in nature. Here in the Sahara I was with other people, but communication was limited to gestures and grimaces.

Solitude, however, seemed to be the last thing these desert dwellers wanted. Whenever we sat down, with the enormous spaces around us, they formed a tight clump. Men and boys would hug, hold hands, or lay their heads in one another's laps. During the frigid nights such clustering helped share body warmth. But the nomads huddled together even in the heat of noon, seeking human contact in the desolation.

I was only a visitor though, and I welcomed the desert's magnificent isolation. When we rested or slept, I looked for a dune far from the camp, where I could bask in the unique, complete silence. Often, the only sound I heard in the cold night was the clattering of my teeth.

One day we passed an area littered with what looked like ancient stone implements—axes, scrapers, spear points, mortars and pestles. We had apparently come upon an old dwelling site that a sandstorm had uncovered. When I collected some of the nicer pieces, the nomads laughed. To them, these relics were just rubbish.

I carried the artifacts in my hands, since my luggage was strapped to the top of my camel, and I could hardly stop the whole caravan just to pack some "rocks." After carrying the load up and down dunes and over a couple of plains, I started to jettison my finds one by one. Eventually I dropped all of them except for one small arrowhead which I placed in my mouth to keep the saliva coming, so my tongue wouldn't stick to the roof of my mouth.

On the ninth day of our trip, as we approached a long valley between two giant dunes, half a dozen women and girls came running toward us, ululating wildly. Dah and the boys waved from their camels. When the welcomers got close enough to see me, they ran as fast as they could to hide behind a hill of sand. I combed my hair, hoping to make my appearance as benign as possible, while the boys of the caravan laughed at their kinfolk's shyness. Probably none of their mothers and sisters had ever left the desert, even to visit Timbuktu, and in all likelihood none of them had seen a European before.

Dah came back to untie two of the camels from the caravan, gave the younger kids instructions, and disappeared behind the dunes again with the camels in tow. I followed him up the valley with Hadji and Mohammed. It was at least an hour before we came to the two tents that made up Dah's camp, one for the women (Dah's wife, the widow of his dead brother, an unmarried cousin, and three girls) and the other for the men (Dah, Hadji, Mohammed, and four or so younger sons). This was the whole population of the camp.

The tents were quite large, covering an area of about thirty feet square, but looked nothing like any tent I'd ever seen before coming to the Sahara. One was made of goat hides and leather, the other of little strips of cotton, sewn together to make a sheet, both propped up with crude wooden poles stuck in the sand. The result was simply a large shelter against the sun, with open sides and the bare ground for a floor, the family's belongings scattered in the sand. During cold nights, I later found out, the windward flaps were let down to create a shield against the weather. Between the two tents a nanny goat and her kid were tied to a stake, with some desert straw nearby for them to eat.

But for now the site was completely deserted. Where were the women and children? They seemed to have disappeared as mysteriously as they'd known when to come meet us.

I felt out of place for the first time. I had been perfectly comfortable traveling with the men of the caravan, but meeting their families was another matter. What if they refused to eat in my presence? Before coming to the Sahara, I had read about the striking difference between the Tuaregs and the Moorish Arabs in their contact with outsiders. The books said that entering a Tuareg tent with women and children is hardly ever a problem. The lean, tall Tuareg men always keep their faces obscured, but not the women. They have much greater autonomy than do the Moorish women, who almost always wear veils and would never allow a man outside the family to touch them.

Throughout our trek, women had been nearly invisible. Whenever we had approached a camp, the wives and daughters had invariably disappeared inside at the first sound of our approach. On those rare occasions when I caught a glimpse of some female, her face, hands, and legs, every bit of skin was totally hidden in threadbare black or indigo cotton wraps.

Dah returned on foot, and we began to unload the caravan. I used hand gestures to communicate my uneasiness to him, and he cheerfully motioned back not to worry. About a hundred meters from their tents, Hadji, Mohammed, and he set up a shelter for me, a blanket held up by a couple of sticks. We put my luggage underneath it, and prepared *crème*.

"Make yourself comfortable," Dah conveyed to me with a combination of gestures and unintelligible words. "This might take some time, but everything will be all right." At least that's what I hoped he meant.

We had tea, and Dah brought out a bag of dates, to which I added some candies I had bought in Timbuktu. Then he and his sons mounted their camels to have another try at coaxing their relatives back to camp.

When my friends had gone, I lay down in the crude shelter, tired, elated, dirty, and worried. I was somewhere I could never have found on any map, in the middle of the Sahara. Occasionally I had checked our course with my compass, but I knew only that we had traveled in a general northwesterly direction from Timbuktu for nine days. I had visions of cold beer.

The silence in the empty camp became oppressive. There was no sound at all. No bird, no wind rush, no voices, nothing. All I could hear was the noise of my own body: my heartbeat, my breathing, the slight rumbling in my stomach as the *crème* wound its way through my guts.

I pulled out Hemingway to pass the time, but I didn't feel

like reading about the Great White Hunter's exploits. I rummaged about until I found my issue of *Gentleman's Quarterly,* and familiarized myself with current theories on the proper length of cuff to be left showing beneath a man's jacket sleeve.

Around sundown my hosts appeared over the dunes. Dah was walking with three women, all shrouded in black. Behind them rode Hadji and Mohammed, each with a child or two beside him on his camel. I had no notion of the proper etiquette, and *Gentleman's Quarterly* certainly had nothing to offer.

Sheepishly I stood next to my shelter with a handful of candies. I tried on my biggest, most obvious smile. Hadji and Mohammed unloaded the children, who scurried behind their mothers to hide. The women, in turn, tried to hide behind the men. We all stood in a line, silhouetted against the setting sun in the middle of the desert, each waiting for someone else to take the first step. It was awkward, to say the least.

I held out the candy. Nobody moved. Then Dah spoke a few words to the women, and one of them came toward me. Her veil parted a little, uncovering one eye and a bit of her nose. A hand emerged from the worn black cloak. Red and black designs covered the fingers and palm.

Should I shake hands? I asked myself. I decided to play it safe. I dropped the candies into her hand, hoping I hadn't broken any taboo.

The candy disappeared in the folds of her garment. She turned and walked around the men and the other women to the children. She put a bonbon in her mouth and sucked it tentatively. Then she gave one to each child. When she turned back to me her face was a little bit more uncovered, and it bore a big, wide, friendly smile.

After that everything became easy. I went to my bag to get more sweets, and the kids ventured close enough to receive them. The two other women went to their tents to prepare dinner, while Dah's wife crouched in the sand beside the kids. She looked at me with no trace of embarrassment, her face uncovered. I could not make out the outlines of her body beneath the loose black cloak—which is why, of course, women wear it—but as the evening breeze blew her wrap against her body I could tell she was on the chubby side. A fat wife, the nomads say, proves her husband a good provider. Occasionally I'd catch a glimpse of her hand, which apart from the intricate red and black designs, was pale unsunburned white. She had the most beautiful teeth I'd seen in a long time.

Dah regaled his family with tales of our journey, undoubtedly telling about his misadventures with the crazy foreigner. One of the other women brought over a tray of embers and a kettle of tea. The children huddled close to their parents, their big eyes on me, still not quite sure what to make of this stranger.

That evening Dah butchered a goat with my Swiss army knife. He had absolutely no idea of the uses for the screwdriver, the bottle-opener, or the corkscrew, but he was fascinated by each shiny metal tool and demonstrated it for the whole family.

Every part of the goat was eagerly devoured, with the exception of the gallbladder, which tastes too vile. None of the intestines were washed; Dah simply cut them into little pieces and boiled them in dirty water. The head, with the hide still on, was thrown into the embers and became a treat for the children even before the rest of us started to eat. It was startling to see the eldest girl dig out the eyes for her smaller siblings, and then pick the roasted gums from between the yellow teeth. Dah cracked open the skull for them with the handle of his knife, so they could take turns sucking out the brain.

To my great dismay I found that, as the honored guest, I was entitled to all the choicest organs. That meant the heart, the pancreas, the spleen, the guts, and other such delicacies. Regular meat was for the "lesser" people, and never before had I wished so much to be "lesser." Dah put these tidbits directly into my mouth, so I dared not refuse. I swallowed bravely, mostly without chewing. In the interest of politeness I kept myself from grimacing, although my forced smile only encouraged Dah to feed me more.

Women and men sat in different circles, but only a few feet apart. The women, quite unlike the women in Timbuktu, ate exactly the same things as the men. The children, though, got only the bones, and these only after they'd been chewed on by the adults. Since the butchering and roasting were done in the sand, every bite had a good portion of grit. It would take me a while to get used to that.

I slept well that night. I had a shelter, but more important, I had a blanket. Dah had lost mine during my involuntary stay in Timbuktu, and since I rejoined the caravan, I'd been sleeping with my ski jacket under my *boubou*. He had offered me his own coverlet, but I couldn't bring myself to leave him with only his threadbare jellaba and billowing knee-length pants. Tonight, with one blanket to shield me from the wind and another to wrap around my body, I slept blissfully.

The nomads' household arose just before dawn. When I woke up at 5 A.M., children were already pacing anxiously around my lean-to. They smiled at me, but stayed at a safe distance. I smiled back at them, not sure what was expected of me. Should I go over to the main tent? Should I offer to help? Help do what? I had no idea what a day in this camp was like. I didn't want to be a burden, but I did not know what I could possibly contribute.

When in doubt, drink tea. Dah came over with his little kettle and two glasses. He cleaned them by rubbing them with handfuls of sand. Then his wife brought over a bowl of dried camel meat soaked in hot water, and a little girl carried a bowl of curdled camel milk. Breakfast was served.

We stayed for a week or so with the camel driver's family and a trickle of other nomads who wandered into camp. I soon found I wasn't mere dead weight after all: Dah distributed the bulk of the provisions I'd been forced to buy to the impoverished members of his clan, and I began to lose the bitter taste of having been royally conned in Timbuktu.

But besides being a cash cow, I was also a source of amusement for everyone in camp. Dah's children had no television to keep them occupied, but now they had me. The children scampered around like little dervishes with my miniature flashlight. Nobody gave even the slightest indication that they expected me to give any of my gadgets away, but no one had any qualms about borrowing them at any time. I did make one restriction, when the little ones wanted to clean their nails with my toothbrush. I later found out that they kept their own teeth so white by rubbing them with a piece of wood for hours.

One morning, one of the boys, who had been watching me as I tidied up my shelter, suddenly ran off and returned with a pointed stick. He plunged the spear a few times into the sand near my feet and pulled out a three-foot-long horned viper. Apparently, he'd seen the snake's snout and eyes sticking out of the earth. With a grin, he scampered away carrying his prize.

Another time, while we were all sitting around the fire after a meal, Dah got up and approached me with a sandal in his hand. He smiled as he nonchalantly squashed a scorpion on my blanket, then shuffled back to his seat by the fire. On yet another occasion, what had looked to me like an ordinary beetle before Dah had mashed it to an unrecognizable pulp with his sandal made everyone in the camp jump up with alarm. I wondered what kind of insect could be so deadly that the nomads feared it more than a scorpion.

My hosts seemed to take their religious duties rather casually. While dinner was cooking, or during any other pause in activities which happened to come near one of the five obligatory hours, they'd turn toward Mecca and pray. Should an interesting conversation start nearby while they were at it, they'd simply turn around to take part and resume their supplications at the next convenient break. One man, apparently a friend of Dah's who had joined our caravan with his son and a few camels, liked to have his little boy walk on his back as he knelt and bowed in the sand.

The whole time I was with the nomads, in both caravan and camp, I did not see a single fight among them. By any material standard, they are extremely poor. The small children go naked, the men wear only jellabas made of cheap cotton, and the women are draped in dark indigo-colored cloth that stains their skin purple. Apart from their tents and their animals, a few iron bowls and utensils are their only belongings. I do not know whether their social harmony is despite their deprivation or because of it.

All the nomads, men, women, and children alike, start to doodle in the sand whenever they sit down. First they prepare an area in front of their folded legs, smoothing it out with a hand and removing any twigs or camel turds, like a teacher wiping a blackboard before class. When they talk, they make patterns of dots in the sand, points for emphasis. When they get up they wipe the designs away, as if they want to leave behind no trace of their ideas.

When the nomad men go away, whether to the mines or to Timbuktu, they make a camp for their women in an area with enough pasture to maintain the livestock until their return. Since such abundant pasture is seldom in the vicinity of a well, at least one man has to stay behind to go get water—it would not be "proper" for a woman to leave camp unescorted and mingle with strangers at a well.

Much of the nomads' life is geared toward obtaining basic necessities. The men transport salt from the mines to earn the means of purchasing sugar, tea, grain, kettles, utensils, fabric for clothing, and the rare ancient rifle. In good times, when the rains permit herds to multiply, they also breed goats for barter. But since all these transactions require constant travel, the men and women spend much of their lives apart.

The nomads keep male and female camels separate most of the time as well. Only male camels are taken on caravans (to avoid disruptions from mating or fights). The females are left with the caravaners' wives and children back in camp. The she-camels are permitted to wander freely, but the herders have a rather ingenious system to bring them in from the desert for milking: As long as a she-camel is in milk, the owner ties a small basket over her udder so the calves cannot drink. As the pressure of the milk builds up, the mother becomes desperate for relief. She has learned that her master will take the basket off, so she comes to camp. The herder takes what milk he needs for his own family, and then lets the hungry calves suckle at will. When the men are away on a caravan, these tasks are performed by the women and the bigger children.

The sounds that a small group of female camels can make defy description. There is mooing, gurgling, burping, slurping, moaning, groaning, uuuhing, and aaahing. One night when they were particularly vocal, Dah managed to make me understand that such volubility is generally taken to mean that rain will come soon. We were five months away from the rainy season—and even then it only rains every couple of years, and just in certain parts of the desert—but the next morning the sky did turn overcast and gloomy.

After a week we broke camp and moved on, the women and children coming with us this time. Hadji and Mohammed loaded all our belongings onto the camels. One of the animals carried Ouija (Dah's wife) and their three youngest children, all on a platform made out of bundles. The other two women shared another camel, which also carried the tent poles on either side. The three adults, the two bigger boys, and I walked. Where to? I had no idea. According to the compass we were headed east.

It is an average of five days between wells, so when we did arrive at one we all drank eagerly. Each well is a hole about 6 feet wide, probably about 150 feet deep, surrounded by a mound of camel shit built up over hundreds of years. Whenever a nomad draws up one of the leaky leather bags of water, he invariably sloshes most of it onto the ground, where it softens the caked manure. This slurry mess of camel dung, goat dung, and water naturally runs back into the well—to be drawn up again in the next load and consumed by humans and animals.

I quickly learned to shut my eyes whenever I drank the brown water, though the taste still made me wince. I had decided before coming to the Sahara that I would eat and drink exactly the same things as the local people, to get a better feel for the way they lived. I'd been told that the chances of picking up a disease in the desert were slight, and at any rate, the way I was traveling, I could never have avoided exposure to whatever microbes might be around.

Since I could not communicate verbally, I had to use all my senses to absorb whatever information I could. When the women pulled the stakes from their tents and rolled up their blankets, when the goatskins were emptied of water without anybody going to a well to refill them, I knew it was time to pack my bag. Once we had loaded up, we would stop at the next well to water the animals and fill up every goatskin for the long march ahead.

I had been wondering what the camels and goats ate, since I could rarely find even a single piece of straw to start a fire. One day we were joined by a caravan of about 150 camels, and I learned the answer. The convoy was carrying huge bundles of *halfa,* a type of straw that grows wild in some parts of the desert and is the main fodder for pack animals. The nomads bundle the *halfa* to take with them on trips to the salt mines. Most of that 800-kilometer haul is through areas completely devoid of vegetation, so as they go, they leave caches of feed in the desert to keep the camels from starving on the way home.

Halfa is virtually indispensable to the desert dwellers. Except for a few camel-hide strings used to bind salt bars, all the rope that the nomads use is made from *halfa*. The straw is also used as kindling to start camel-dung fires and as padding pouches to cushion the camels' flanks against the load of baggage.

Whenever we passed an area where *halfa* grew, we loaded sheaves of it onto our camels. At night, while sitting around the fire, the whole family would make ropes of it by twisting the strands between their hands in rapid, complicated motions. It looked very easy, but every time I tried, I ended up with a handful of crumbled straw. I made up for my failure, though, by teaching the children how to skip rope with the lengths they'd just made. With all the rope they used in their daily lives, it was odd that they'd never discovered the game for themselves. I was pretty good at it—the legacy of a youthful ambition of becoming a champion boxer.

We got belly aches laughing at the children and occasional visitors from other camps, as they entangled themselves in the rough straw ropes trying to emulate my virtuosity. I also taught them the game of tag, which we often played when the daily chores were done. On moonless nights, with only starlight for illumination, tag soon became more like blind-man's bluff.

By this point I had finished Hemingway's *Green Hills of Africa,* and I knew from *Gentleman's Quarterly* which color necktie I should wear this year. And then my roll of toilet paper ran out. Since I was trying to live like the nomads I should have made do with sand, but I didn't feel quite ready for that, so either Papa or the *Quarterly* would have to serve.

There were not really any ties around, regardless of color, but one never knew when such information might come in handy. And *Green Hills* did have so many repetitive passages about the Great White Hunter measuring antlers or molars or tails or whatever, just to discard the beast because Carl had already gotten a bigger one . . .

Next time the need arose, out came the part about the sable antelope whose horns had been just a bit too short.

I'd noticed that while children were allowed to run around completely naked, after the age of six or seven the nomads were very prudish about displaying their bodies. As part of my Saharan training, I learned to abide by this code of modesty. Like the others, I learned to heed the call of nature behind the nearest dune, which wasn't always so near. I developed prodigious bladder control.

The tables were turned, however, the day I taught the older children how to stand on their heads. Dressed in trousers and not giving much of a damn about who might see what anyway, I could balance upside-down quite easily. But the kids were a ludicrous spectacle. They could stand on their heads, or they could keep their flowing robes discreetly covering their privates, but they could hardly do both at the same time. We couldn't stop laughing even when a scorpion crawled out of my bedroll.

Often I tried to ask Dah when we'd meet up with a salt caravan, because I wanted to see one, and I knew he would not be traveling to the salt mines on account of me. Each time he answered with a gesture that meant, "Soon, soon." Then one day a caravan came to our camp. Some of the animals were loaded with bales of *halfa,* so they were obviously headed for Taoudenni. I hastily got my gear together, Dah loaded my camel and his with a few bags of grain, sugar, and tea, and the two of us took off with them. Hadji and Mohammed stayed with the women and children, to help them find good pasture for the female camels and goats.

Much later, I learned that this meeting had been arranged. Nobody goes to Taoudenni without a recognized guide, one who has the "gift" of navigation. Such men are few and far between. Dah had the gift and was originally supposed to direct this caravan, but since I was barred from Taoudenni, and he was my chaperone, the salt traders would have to find somebody else. They were going to a village where they could find a caravan with a guide, and the convoy's leader (a nephew of Dah's, I think) invited us to tag along. At this point I had no idea where we were going, or why, but I decided to go as far with them as I could.

One afternoon we came upon the tracks of a gazelle. Dah halted and said something to one of the other men, who took off on his camel at a trot. I couldn't quite imagine how he was going to catch up with one of these incredibly swift animals, which seemed to float effortlessly over the ground in the distance. Dah explained to me with signs that he had seen from the tracks that this particular animal had been wounded and was dragging one hoof. A few hours later the man joined us again, with the gazelle over his shoulder. He had no gun, no bow, and no spear. His only weapon was the short dagger which all the nomads carry tucked into their belts. Dah gutted the animal right away. He cut the liver, heart, intestines, and other organs into little pieces and handed them around. We ate all the innards right there, unwashed and uncooked. The liver, the heart, and even the lungs weren't so bad, but the intestines—still full of what would soon have become gazelle droppings—gave me the creeps. I didn't chew, just swallowed as fast as I could.

On the fifth day since leaving his family, Dah pointed out a speck in a whole mess of dunes. He said a name, but it was one I didn't understand. We kept on walking, and the speck disappeared behind other dunes. Occasionally it would reappear. We searched for a passage through the mountains of sand; a loaded camel cannot go straight up a steep incline, or the cargo on its back will shift and send the animal crashing down.

When we finally reached the speck, I saw that it was an actual village, complete with houses, the first permanent settlement I'd seen in what seemed like several lifetimes.

But what a horrible place. Why anybody would have built

a village here was beyond me. No vegetation, no shade, just sand and rubble. No beer, either. And ravenous swarms of black flies buzzing over the inhabitants in their tattered rags, and over the camels and goats and salt bars.

Surely this was hell on earth, I thought. I turned to Dah, who seemed to read my mind.

"Araouane," he said.

NOWHERE SPELLED BACKWARD

A man's typical day here goes as follows: He gets up before sunrise to pray, then drinks his three glasses of tea—if he has any. This takes about two hours. After tea, he joins the other men on one of the dunes and sits and watches to see if some food might come out of the sand or appear out of the desert. Around noon, he goes back home to drink baobab powder and pounded millet mixed with water—if he has any. Then he sleeps for about an hour, and if he's lucky, when he wakes up, there might be food for lunch. In midafternoon, he goes to the little mosque to pray for food. In the evening, more tea on an empty belly.

Women and children have a busier routine. Unlike the nomads, they eat apart from the men. They spend the whole day looking for crickets, lizards, and locusts, which are almost their only source of food. Sometimes they get lucky, and spot a flock of tired birds flying overhead. They dash to set up camel saddles, blankets, buckets, anything to create a bit of shade. The birds see the cool bits of shadow, swoop down to rest for a few minutes, and the women and children slap at the scrawny things with damp rags until they are dead.

Dah's nephew and the other caravan members stayed in Araouane only a few days, just long enough to find a guide to the salt mines (that is, to tag along with a passing caravan that already had one). Dah and I stayed about a week. During that time the townspeople's only regular activity seemed to be watching me. Every night the house where I lodged was overflowing with visitors. Each person hopefully brought his own kettle and glass. Dah supplied them all with tea and sugar. They'd sit—or, once the room got too crowded, stand—for hours, sipping their tea and staring silently at the odd newcomer.

Sometimes one of them would bring me an egg—quite an extravagant gift, since the village had only a handful of bony chickens. I had no frying pan, so I had to suck out the insides raw.

The only person I could talk with was a young Moor, who still remembered a bit of French he'd picked up in Qur'an school in Timbuktu. He told me a few stories of Araouane's illustrious past, but it was only when I returned to the libraries of the West, that I learned just how proud a pedigree this little village had once had.

Nobody quite knows how the town got its name. Some people say it comes from an archaic Tamacheq (the language of the Tuareg) phrase meaning "the place where one needs many cords to pull water from the wells." By pure coincidence, it is pronounced much like Erehwon, the utopia imagined by the nineteenth-century writer Samuel Butler, who arrived at the name simply by spelling "nowhere" backward.

According to local legend, it was settlers from Araouane who founded Timbuktu. Tuareg from the desert (the story goes) traveled to the banks of the River Niger during the dry season. There they established a camp where they could leave some slaves, grain, and utensils under the protection of an old woman named Buktu. When the nomads turned to settled life, they built their city, Timbuktu, "at the place of Buktu."

As Timbuktu grew into a seat of the Sorai Empire, Araouane became a center of trans-Sahara trade. Araouane had the only deep wells in that part of the Majabat al-Khoubr, the Empty Quarter, where water was as valuable as gold. From the twelfth century on, great caravans of as many as ten thousand camels stopped to drink at Araouane. All the trade from the fabled Gold and Ivory Coasts to Mediterranean Europe, the Middle East, and Northern Africa had to cross the Sahara. Much of it had no choice but to pass through Araouane.

From the north came horses, glass, coral, and copper utensils, mercury, dates, leather, seashells, wheat, raisins, and salt, weapons, resinous woods, Egyptian linens, Indian cottons, and fine carpets from Persia or the Orient. From the south came pelts, perfumes and spices, molasses, kola nuts, and ostrich feathers, ivory, gold, and slaves.

It was Araouane that linked the great North African centers of Tripoli, Fez, and Marrakech with the great Niger River depots of Timbuktu, Gao, and Djénné.

Araouane first entered recorded history as a real town around the fourteenth century, as chronicled by the Islamic holy man, Tuareg Agmed ag Adda. By 1470 the town was important enough to possess both a judge and an imam. A sixteenth-century traveler and explorer named Leo Africanus described Araouane as the granary of all the surrounding Berber tribes, who by that time had conquered the area from the Songhaï and the Tuareg.

At its height, Araouane was far more than a trading hub. It was a center of culture and faith. During the Middle Ages, Araouane was home to no fewer than 300 Muslim saints. Timbuktu, in a classic case of one-upmanship, then claimed 333.

But with the growth of shipping in the sixteenth and seventeenth centuries, Araouane's importance began to dwindle. Gold and ivory were siphoned away to Europe rather than the Middle East, so the merchants of the desert turned more and more to slaves. As the great empires of Dahomey and Timbuktu slid into decline, so did Araouane.

In 1828 the French explorer René Caillié described the village:

> *The streets are wider than the ones in Timbuktu—*
> *and clean. . . . The stores are very narrow; there are*
> *about five hundred houses, all not very well built,*

each with an average of six inhabitants, including
slaves. . . . The Moors search for their camels in six
day intervals to water them at the wells, in the vicin-
ity of the town, which are about sixty ordinary foot
steps deep.

But Caillié called it the most miserable place he had ever
seen, with unfriendly people and more flies than anywhere
else. And he had seen plenty of miserable places in his wan-
derings.

A mere twenty-five years later, another European reported
that the population, estimated by Caillié at 3,000, had
dropped to 1,500 souls. He also noted that the black Sorai
had long since replaced Moors as the most numerous ethnic
group.

With the death of the slave trade, Araouane began to die as
well. Commerce across the desert dwindled. With no slaves
to buy, merchants had little use for the salt bars that had
served as currency for centuries. Once exchanged pound for
pound with gold, salt now became almost valueless. The
grand caravans grew fewer and smaller.

At the end of the nineteenth century, a German traveler
named Oscar Lenz passed through once-famed Araouane. "A
horrible situation," he wrote. "Not one tree, sand and flies
everywhere."

When a census was taken in 1931, Araouane was found to
have only 255 inhabitants. The French colonial officer who
conducted the survey lamented, "Soon all that will remain
will be a memory of a town, once prosperous and a
renowned center of learning."

On March 22, 1991, during my third year in Araouane, I made
my own census of the village. I found thirty-one inhabited
houses, twenty-two abandoned ones waiting to be reclaimed
by the desert, and three mosques. There were only 145 people left: forty-five women, thirty-four men, and sixty-six chil-

dren. One hundred and eight of the residents were black Sorai, thirty-seven were Arab Moors. For as long as any of the
villagers remembered, no Tuareg had lived in Araouane.

As for Araouane's unruly neighbors, the Tuareg too had a
long and proud history. They had once been the undisputed
masters of the region of the Sahara that covers what is now
southern Algeria, Mali, Niger, Libya, and Burkina Faso. It is
an area greater than the American southwest from Texas to
California, but has fewer than 1 million inhabitants. The Tuaregs once ruled this whole territory, roaming the desert with
their flocks of camels and goats while they kept black slaves
on fertile oases to tend their date palm groves. They took
other blacks with them into the desert stretches as household
servants; their blacksmiths (Tamacheqs, mostly from the
black Bella tribe) were famous for their beautifully crafted
daggers, swords, and jewelry.

The Tuaregs got their slaves by raiding black villages in the
south of the Sahara and earned money by selling their extra
captives in the north. This practice lasted from the Middle
Ages right up to French colonial days. After a few years of
fierce resistance to the French newcomers, toward the end of
the nineteenth century, the Tuaregs were reduced to simple
herders once again. Some settled down in the towns of the
southern Sahel region.

The French had required all families living in their domains
to send at least one son to government-sponsored school.
The Tuaregs and Moors were suspicious, assuming that the
Europeans wanted to corrupt the minds of their children and
perhaps even convert them to Christianity. As a ruse, they
would send the sons of their black servants, and keep their
own children at home.

This practice, paradoxically, lost the Tuaregs whatever
chance they may have had to regain control of their territory
once the colonists left. When West Africa gained its independence from France in the 1960s, most members of the former
Tuareg and Moor ruling clans were unable to speak French
and often illiterate even in their own languages. Due to the

large number of tribal tongues spoken in the region, all of the new nations of former French West Africa adopted French as their official language for business and official purposes. Tamacheq (the language of the Tuaregs) and Hassania (that of the Moors) were so different from classical Arabic that even the Muslim imams of other tribes couldn't understand them, and neither group could hope that theirs would become a regional lingua franca. French was the language of power, and most of the people who had learned French in school were black. The former rulers now became an underclass, governed by black (mostly from the Bambara tribe) bureaucrats from the south.

Disastrous droughts of the 1970s and 1980s forced even more Tuaregs and Moors out of the desert, which had become too inhospitable even for their hardy herds. The once-proud nomads now became pitiful refugees begging in the Sahel. Predictably, there was little love lost on them by the black administration. Aid from international organizations, earmarked for the Tuaregs and Moors, often "disappeared" before it could reach the refugees.

The nomads began disappearing as well—back into the desert, with stolen all-terrain vehicles and weapons, emerging only to raid villages and army posts. The troops of the south were strangers to the desert and could not follow the outlaws without Tuaregs or Moors to guide them. So the revolt simmered on, embattled but never wholly crushed.

That is how things stood when I arrived.

Throughout its history, Araouane had always been able to rely on sustenance from outside. Its wealth had been its wells, and water had always bought all of life's necessities. The caravan traders had been happy to sell their weak or unneeded camels to the Araouanites for meat, and the passage of so many beasts had always left plenty of dung for fuel. Araouane had prospered as an oasis, albeit one without vegetation, but such an oasis can prosper only if it has visitors.

Without the cross-desert commerce in slaves, salt, and gold, visitors stopped coming to Araouane.

After I'd been in town for a week or so, I found myself wondering how I might help the inhabitants get back on their feet. It wasn't any highblown sense of charity that had me entertaining such thoughts. It was a challenge, a bit of excitement, something to pass the time. Anybody can visit an exotic place, but to really change it for the better—now *that* was an adventure. As it turned out, it was a project that would come to dominate my life, but in the beginning I just fell into it.

The townspeople could never again live solely off trade, I reasoned, since the huge camel caravans were a thing of the past. In order to survive, they would have to become self-sufficient. That meant Araouane would have to grow its own food. I decided that what the town needed was a garden.

Using the Moorish boy to translate, I recruited a few of the children to help me collect camel bones. There were plenty of stripped skeletons lying around, and the bones could be used to make some sort of enclosure for seed beds. Of course, little can grow in pure sand, so we would have to create some soil. We collected a huge pile of hard, dry camel dung, and started crushing it into crumbly pieces. By the end of the afternoon the whole village had gathered to watch, and when a critical mass of the kids knelt down to start crushing shit alongside me, everybody got in on the act.

I told them to water the manure-bed daily, so that it could start to rot and turn into decent soil. And I promised to arrange for vegetable seeds to be sent by the next caravan when I returned to Timbuktu in a week or so. At first I planned only to help set up a modest little patch, send on some seeds, and let the villagers take care of the rest. But as I gave the plan more thought, I realized how unrealistic it was to expect people with no concept of agriculture to start farming without guidance. I decided that I would come back and help them, at least for a little while.

• • •

That night, I lay awake scribbling notes in my diary:

THINGS I WOULD NEED IF I CAME BACK TO ARAOUANE:
1. An all-terrain vehicle, so as not to be dependent on camel convoys for travel. Would need a car equipped for desert driving, something big enough to sleep and cook in.
2. Seeds from Europe, specially selected for this environment.
3. Sprinkling cans—we could also use regular cans with little holes punched into the bottom.
4. Many, many straw mats from Timbuktu to make wind, sand, and sun shelters.
5. A lot of rope and string to hold things together.
6. Plenty of sticks for supporting the mats and growing stalks; there is no wood here at all.
7. Lots and lots of buckets for carrying water.
8. Patience, patience, patience.

I woke to the sound of flies buzzing in the house and, outside, a raging sandstorm. The insects were seeking shelter, to avoid being blown west toward Mauritania. I could hardly decide whether the pests or the storm was more unpleasant.

I poked my head out the door, and the swirling sand was so thick I could barely see the ground. The sun was merely a slightly brighter patch in an ugly yellow sky. There was no horizon at all, just earth and air melting into a single murky mess.

I walked out to the nearest dune, about 200 feet from town, to be alone with the storm. Plumes of sand raced in wild eddies, and the earth seemed to move like sluggish lava. It was strangely silent, as if the blasting sand drowned out all other sounds. There were no leaves to rustle, of course, no doors or windowpanes to rattle. But when I put my ear to the ground, the moving sand tinkled like wind chimes in a stiff breeze.

Eventually I went back inside, but I was ready to leave. I wanted to get away from the fly-infested house, away from people staring at me constantly, away to the wide, uninhabited desert. Dah noticed and seemed relieved. He too was anxious to get back on the road.

During a slight lull in the storm, we made out two camels loaded with salt wandering about unattended. There were no caravans in town that we knew of, so they must have gotten lost. This might mean there were stranded travelers nearby as well, in danger of dying from lack of water. But nobody went out to look for stragglers. It was impossible to know where to look, and members of the search party would risk getting lost themselves. Without the stars to guide them, they might never find their way back to Araouane. We built a signal fire, burning old goat skins too worn to carry water. But the glow illuminated nothing but the plumes of sand.

Dah and I could not delay our return to Timbuktu indefinitely, so we loaded the animals in the furious storm. The sand was so thick I could barely keep my eyes open. Even the camels were frightened and made none of their usual grunts of protest as we fitted them for the journey.

For the next two days I saw barely a glimmer of sunlight. My ears and nose were always full of sand, not to mention my clothes and baggage. I wanted to take some photographs, but trying to load film would have ruined the camera in an instant.

When I woke up on the morning after our first day of travel, I found that a small dune had formed around me during the night. My body was covered by at least a foot of sand. I dug for my shoes, flashlight, and flask, which still had some of Araouane's delicious clear water. After a few gulps, I understood why Araouane's wells were famous throughout the Sahara.

All day I followed Dah closely, fearing that if I lost sight of him even briefly, I'd be stranded in the waste. But as we trudged on, I began to feel we were headed in the wrong direction. I became more convinced of it as the afternoon wore on. Finally I took out my compass to check and discovered we were dead on course. I had a burning desire to find out what guided my friend, that man of the desert. We had a lot of time together, only our two camels to care for, no water to draw (we'd brought enough from Araouane), so our communication had become a bit easier. We had reached the level of two little boys from different countries who just jabber away in different languages, somehow managing to convey rough ideas with the help of pantomime and facial expression. I sometimes felt we understood each other by telepathy.

"How do you know which way we have to go?" I asked Dah when we stopped to rest.

Easy, he said, shrugging and pointing to the sky. All you do is keep the North Star over your right shoulder.

But it was still daytime, and in the storm neither moon nor stars were visible even at night.

"How do you know where the North Star is?" I asked.

No problem, he indicated, sticking out his finger.

Again I took out my compass, and its needle pointed in exactly the same direction.

The Sahara is full of men like Dah, who infallibly set their courses without any modern orienteering gadgets. Stories of the most famous guides are a desert staple. It is said that one renowned pathfinder was leading his caravan across a flat expanse of terrain that stretched for hundreds of miles without a single discernible landmark. "At noon tomorrow," he announced casually, "we shall rest at the spot where I lost my pipe during last year's expedition." Sure enough, just after the midday meal, he scratched the sand in front of the fire, dug up his pipe, and had himself a peaceful smoke.

• • •

When we got up at dawn the next day, the wind had died down to a light breeze. In its place, though, was a big black cloud of flies. Sandstorms may be dangerous, but at least they blow the flies away. I had grown perfectly comfortable riding on my camel for hours or walking in the hot sand, but I still could not deal with the desert insects. They were everywhere. Flies were the worst—every living thing was covered with whining swarms. Moreover, whenever I'd sit down, an army of huge silver ants would crawl all over me, fast as phantoms and burning like firebugs. Anytime I happened to be near a bush or shrub, I'd have to fight off spiders and big zebra-striped insects I had never seen before. At night, the *gang-gangs* would come, large bulldozer bugs that spend their lives trying to burrow under anything in their path. They aren't harmful, but every time one of them digs beneath your body, you have to jump up to make sure it isn't a scorpion.

That afternoon we came to a major watering hole. The fresh water was welcome, but the well was even more repulsive than most. It stood in the middle of a true mountain of camel, goat, donkey, and sheep shit. There were a few dead, twisted tree trunks, around which you could stretch a rope to draw water. An old man, wearing a tin *crème* bowl on his head like a knight's helmet, had set up a pulley and was savagely beating his donkey. She was so pregnant she could barely move, but the man forced her to hoist a heavy water-skin up from a depth of at least 50 meters. The scene offended my morals and my sense of hygiene—but I nevertheless shoved goats and camels out of the way to drink lustily from the rusty barrel into which water was poured for beasts and men alike.

Three days past this well, Dah pointed to the sun and pantomimed drawing it down to the horizon in the east, which I easily understood as sunrise. Then he pointed to the camels' feet, to our own feet, and made marching motions. He traced an imaginary arc of the sun across the sky, to the Western horizon, and said, "Timbuktu." We would reach Timbuktu the following evening.

I smiled at him, at the camels, at the sun, nodded my head, and said, "Timbuktu—beer."

Dah grinned back, as he always did when we managed to convey our thoughts to each other. He must have thought "beer" was my language's word for "I am glad." How right he was.

On our last day out, I decided to test the limits of my endurance. I thought I'd gotten used to nomadic life, but I wanted to find out for certain.

We broke camp quite late, because the camels had strayed farther than usual, and it took a good three hours to find them. When we started marching, I deliberately did not put the turban on my head. I left my drinking flask packed away in my luggage. I wore only jeans, a shirt, and sneakers. My goal was to walk all day in the glaring sun without food or water. Someday, I reasoned, it might be useful to know just how far my unassisted range of survival went.

After the customary two hours' walking, Dah got on his camel and waited for me to follow suit. I pointed at myself and made marching motions. Dah pointed at my behind and raised his eyebrows. I shook my head and gave a hop and a skip. He wiped nonexistent sweat from his brow, to say it was too hot to be jumping around unnecessarily. I pounded my chest like Tarzan, to reply that I was tough. He offered me a water flask, but I refused to take it.

"After all this time with us," Dah's expression said plainly, "he is still crazy." He rode on, sitting snugly on his mount, while I half-trotted alongside.

The first problem was that my mouth soon dried up completely. I had never seen the nomads drink between rests, but I had always taken periodic gulps of water to keep my tongue hydrated. I had placed a little stone in my mouth to

keep the saliva coming, but after a few hours there was simply no juice left. My tongue stuck to the roof of the mouth,
and I was so thirsty that I barely noticed the harsh sun beating down on my head.

During our midday rest, when I refused to eat or drink,
Dah got really worried. He must have thought I'd finally
taken leave of my senses. He literally tried to force food and
drink on me. The temptation to accept was great, but the desire to complete the test was greater.

Dah was flustered and didn't understand. He tried to pull
or shove me onto my camel—he even made it sit down so I
might mount more easily, something he hadn't done since
our first day together. He looked around at the empty desert
in desperation, as if hoping to find somebody to help him
with this madman. I tried to convey the reasons for my odd
behavior, but our language of signals was inadequate. He
had never found my intake of food and liquid sufficient for
life in the desert, but the idea that any rational person would
turn down sustenance altogether was wholly beyond him.
The nomads wolf down incredible quantities of food, probably because their diet is so erratic. In times of drought it is almost wholly starch, millet from the south, and in times of
plenty is it almost wholly meat from their animals, never any
vegetables or fruit. They must eat great amounts for their
bodies to absorb enough nutrients. I hoped that back in Timbuktu I'd be able to have somebody explain my behavior to
Dah. I didn't want to leave him thinking he'd driven his
charge around the bend.

At nightfall we finally made it into town. Long before we
arrived I caught the horrid stench; my nose had become so
sensitive in the desert's pure air that I could smell Timbuktu
several miles away.

At the Hotel Azalaï I treated myself to a cold beer, a decent
meal, toilet paper, a shower, and then a whole lot more beers.

The following day I looked up Salah Baba, the merchant who
had arranged what must have been the most lucrative cara-

van in living memory. I'd been told that his clan had been one of the most prominent in Araouane, and had escaped deprivation by moving to Timbuktu. Now that he'd become a wealthy man, perhaps Salah Baba would help me save his ancestral village.

When I told him my plan, Salah invited me to meet all the other important Timbuctiens whose families had come from Araouane. On the terrace of his mansion we had a dinner right out of the Arabian Nights. Seated on rich carpets and embroidered cushions, we feasted on the intestines of goats, sheep, cows, and camels stuffed with strangely delicious spices and all sorts of other delicacies I had never seen or tasted before. After so many weeks in the desert, it seemed like a whole other world.

Salah introduced one of his guests to me as the "chief" of Araouane.

I thought it strange for the leader of a village to live a seven-day journey away from it. And that was not the only odd thing I discovered during the evening. Though everyone expressed great pleasure at my wish to help Araouane, and all of them were quite wealthy by local standards, not one man offered anything more than encouraging words. I hadn't expected them to jump on the bandwagon without a bit of hard thought, but I hadn't asked them for very much: just the use of some camels to help me transport the seeds and food supplies that I would pay for up to their village.

Their smiles made me uneasy. If these rich men cared as deeply as they claimed for the village of their birth, why were the people of Araouane living on crickets and lizards? If these pious men could make pilgrimages to the tombs of Araouane's saints, couldn't they bring some food along with them? Didn't their religion preach charity toward the poor?

What little aid they sent was only to members of their own families, and most of these men had no relatives left in the village. The knowledge that Araouane's fate was in the hands of these characters strengthened my resolve to bring help myself, and soon.

"There is no need to inconvenience yourself by going back there," one of the men said.

"Yes," said another, "it must be very hard for a foreigner to live even here in Timbuktu, not to mention out in the desert."

"You can go back to America, and we will handle all the difficulties at this end," said the first man. "You can just send us the money, and we will take care of everything."

Something told me this would not be a good idea. Still, I'd promised the Araouanites that I would send seeds, so I divided my spare money into three equal parts and gave it to the three men who looked most trustworthy.

Later I learned that none of the men had sent even a single seed.

From Timbuktu I traveled in a very crowded bush taxi through the interior delta of the Niger to Mopti, a city from which I'd be able to get to Bamako for a flight out of Mali. It was still dark when we all piled into the poor excuse for a bus, a broken-down 4x4. There were two rows of seats in the back, holding nine grown men, with mounds of luggage piled all the way to the ceiling and high on the roof rack.

Just getting to the *piste,* the desert track out of Timbuktu, took some doing. These tracks seem to be a well-kept secret known only to the initiated. We made our way westward through a maze of little alleys, honking at people sleeping in the middle of the streets who had to get up so we could squeeze by, then veered off into nothingness. How the driver found our "road" in the tangle of criss-crossing tracks is a complete mystery to me. But somehow he did, and we stayed on it for ninety-seven kilometers until we reached the river at Goundam.

While the car was crawling over a large garbage dump, a man in the back noticed that one of the tires was leaking badly. There was, of course, no spare. Analysis: the valve was broken. Solution: make tea.

After we'd all drunk our necessary few glasses, the driver calmly set about repairs. With almost no tools and only garbage to work with, he quickly had the problem solved. He made a new valve casing out of bits from an old inner tube, which he vulcanized in the embers of our campfire.

Along the riverbank the road became quite swampy. All night we drove through marshland, bogging down at least once every hour. Each time we'd all wake up, pile out, and start pushing. By dawn all of us were covered in muddy ooze. My fellow passengers bore it with good humor. They were all Tuaregs from Timbuktu, going to Mopti for provisions. I never saw their faces, as a Tuareg turban covers everything but the eyes. I was fascinated by their speech, which sounded a bit like Swiss German from Aargau or Thurgau. They seemed to be very gentle people—though I would later learn this is not always the case.

We crossed an arm of the Niger in the morning, where we encountered a scene of confusion. We all had to leave the car to lighten it, so it would not get stuck in the muck. A small wooden boat overloaded with passengers was making its way across the water, when a herd of cattle started wading from the opposite bank. At the sight of so many people, the cattle stampeded. The boat capsized, tossing men, women, kids, and baggage into the muddy waters, which excited the cows even more. My Tuareg companions raised their traveling garments over their heads to protect them, revealing the interesting variety of things (and nothings) they wear underneath. With boatmen trying to prevent their fragile craft from being torn apart, passengers splashing about in search of their luggage, and hundreds of panicked cattle running to and fro in the mayhem, everyone ended up laughing hysterically. Everyone, that is, except one rather elegant lady from another vehicle, whose richly embroidered dress now clung to her body like a soggy rag. I learned later that she was the wife of the district commander of the Timbuktu army garrison, but in her clinging robes she looked like the winner of a wet T-shirt contest.

In the incredibly filthy town of Mopti I saw the ultimate destination of the salt bars from the mines at Taoudenni. After being carried by caravan to Timbuktu, the salt is shipped down the Niger on little skiffs called pinnaces. At Mopti the bars are unloaded, and sold in the market for very little money. Nobody has much need for the salt these days; people in the south sometimes use it for cooking, cattle-feed, or even as a wholly spurious panacea, but less and less frequently. It is no longer a major form of currency outside the desert, so its value is declining practically by the month. I saw merchants haggling carelessly over the bars, trying to get rid of the practically useless chunks, and I was reminded of Araouane's plight.

On returning to New York, my first undertaking was to sign up for an intensive course in Arabic. Well, my *first* undertaking was to stuff my belly with sushi and sashimi at my favorite Japanese restaurant, but I was down at Berlitz the next week. I am no scholar of linguistics, and the language map of North Africa is extremely complex; I knew that the tongue of Araouane was not a textbook dialect, but since it was in the middle of the Moorish Sahara, I figured a grounding in Arabic couldn't hurt.

For four months I spent four hours a day, four days a week, closeted away with a very patient tutor. Most evenings and weekends I studied phrases and vocabulary on my own. Not until my return to Africa, however, did I learn that neither Hassania nor Sorai, Araouane's two major languages, bears any meaningful resemblance to classical Arabic. Sorai has no relationship to Arabic at all, apart from some borrowed words for Islamic terms. Hassania is technically an Arabic dialect, but long ago mutated almost beyond recognition from the pure classical form. It is spoken only by nomads in the western Sahara, people generally illiterate and isolated from the rest of the Arab world, without even television or radio.

As a result, French is the only common tongue that Malians have to communicate with each other. All government officials, traders, and educated people speak it fluently. In most regions it is the only language taught in schools. But Araouane had no school and virtually no speakers of French.

Throughout the summer I tried to squeeze a bit of cooperation out of the Malian government. I hoped that since I proposed to use my own money to help their people, the officials might help out in other ways. Foolish hopes. Telexes to Bamako and visits to Mali's ambassador in Washington proved to no avail. The government would not give me any assistance, not even a temporary residency permit. I would have to set up my project on an ordinary tourist visa again.

After my daily language sessions, I'd spend hours poring over maps of Africa and Asia. My plan was to buy an all-terrain vehicle in Europe, load it with seeds and gardening tools, drive to some oasis in Algeria to learn about desert agriculture, proceed to Araouane, get the garden set up, and be on my way. For my return trip, perhaps I'd motor across the continent to Mombasa, ship the car to Bombay, and from there drive up through India and Pakistan into Chinese Turkestan, Mongolia, and Siberia, where I'd try to find a way across the Bering Straits to Alaska, which would practically be home again. Or perhaps I'd try to drive the old Burma Road to Singapore; I was intrigued by the challenge, since much of that route has been officially off limits to foreigners for thirty years. Of course, Singapore was no closer to home than the Sahara, but that would just provide the opportunity for another adventure.

I never imagined that I'd want to make Araouane my home.

In September, six months after leaving Araouane, I was ready to stage my return. I'd decided to use my brother Peter's farm in Switzerland as a base of operations. Peter, like me, had lived on a financial roller coaster: At twenty he'd hiked penniless through Africa, then returned to Europe and painted houses for a living. Eventually he'd gone into housing construction and made a killing. He sold his company before he turned thirty-five and has been a simple sheep farmer ever since.

When I got to the mountains, the car Peter had ordered for me had already arrived. It was a Land-Rover 110 Turbo Diesel, equipped with sand tires, extra fuel tanks, solar panels, a heavy-duty roof rack, reinforced springs, and an electric winch attached to a ship's anchor for pulling us out of soft sand. We hooked it up to a double horse trailer, to haul my heavy load of saplings, soil, buckets, shovels, tarpaulins, sprinkling cans, seeds, rakes, spare tires, spare parts, medicine, food, and lots of jerricans full of water.

The ferry from Marseilles to Algiers did not leave for several days, so I decided to give my new rig a dry run. I drove through the Pyrenees to Spain. The car and trailer took the mountains in stride, but at the Spanish border the Guardia Civil questioned me for hours. The story I told them—that I was practicing to drive through the Sahara with a mercy shipment for a dying village—seemed suspicious to them. Apparently they thought I had something to do with the Basque separatists.

In the port of Algiers my newly learned Arabic turned out to be a great blessing. I saw petty officials make English- or French-speaking tourists unpack each piece of hand luggage. But when I exchanged a few pleasantries in Arabic, they let my strange cargo pass unexamined.

I had written to Salah Baba, telling him of my plans, and the merchant had forwarded my letter to Sidi Boubacar Balli in Bamako. One of the wealthiest men in Mali, Sidi Boubacar was born in Araouane, and his father had made him promise (on his deathbed, no less) not to let his ancestral village die. Sidi Boubacar had done more for Araouane than any of the other self-exiled nobles and had enriched himself in the process. He had used his influence to get aid from international organizations, and had even sent on to Araouane that which he didn't pocket for his own "commission." Sidi Boubacar had written me back, providing me an introduction to Urassell Kheredine, a wealthy Algerian industrialist who (he said) could help our project.

The outskirts of Algiers were a mess of looted stores and burned-out homes, the result of recent food riots against the Socialist government. I'd heard about the upheaval on European radio, but I hadn't realized I'd be right in the middle of it. Army detachments patrolled most street corners, and armored personnel carriers inspected all traffic. Mr. Kheredine owned a large factory in the suburbs, but because of the rioting, he advised me to meet him instead at an oasis of his in the desert.

I learned more about gardening in the next week than I had in my entire life up to that point. Just as important, I learned how to get a job done with whatever tools were at hand. My host's contraption for drying dates was a good example: It was a chamber fashioned out of material from a junkyard—battered car doors, oil drums, corrugated tin sheets, even pots and pans, all held together by wire. But it worked.

By the time I left the oasis, my trailer was laden with young trees. I had the saplings of thirty olive trees, thirty pomegranates, thirty figs, and twenty-five grape vines. I'd also puzzled out which vegetables could survive in the desert. On the advice of Mr. Kheredine I settled on tomatoes, carrots, beets, lettuce, radishes, squash, turnips, onions, watermelons, honeydew melons, and hot peppers. Just before I drove away, my host gave me a 50-kilo sack of his best dates.

"You can eat them on your way across the desert," he said. "After all, this is the traditional food of Sahara wanderers. Just save the pits, and then you can start your own nursery in Araouane."

As I drove down the Algerian highway to Reggane, with the blindingly white desert glimmering on either side, I sang to pass the time. Hour after hour I belted out a repertoire that ranged from Italian opera to Swiss yodels, but I could not actually hear my voice over the roar of the laboring engine. My favorite tune was an old Mac Davis number whose chorus went, "Oh Lord, it's so hard to be humble when you're perfect in every way!"

Mr. Kheredine had worried that I couldn't carry enough water to keep all the saplings alive on the long trip: I worried more about keeping myself alive. I'd driven through the Sahara once before, of course, when I'd covered 15,000 kilometers in twenty days during the Paris–Dakar rally. But this time I was on my own. No navigator; no race committee to set out emergency caches of fuel, water, and food; no rescue party to come get me if I failed to make it past a checkpoint or activated my hazard beacon. This time I had a hopelessly overloaded car, pulling a hopelessly overloaded trailer. What if my water containers burst? What if my diesel tanks sprang a leak? What if my engine died? I did not want to think of such things, so I sang even louder.

The night before I was to reach Ghardaia, 500 miles from Algiers, a town that is renowned as the most beautiful oasis in the Sahara, I ran into a sandstorm. Visibility instantly dropped almost to zero. The car lights reflected only racing swirls of sand. The headwind slowed me to a crawl, and many times I lost the asphalt entirely. Everything in the car was soon covered with a fine layer of dust. Grainy particles clogged my ears, eyes, and mouth. But at least the storm interrupted the monotony of the voyage for a few hours.

In Ghardaia I bought sand ladders in case the car got stuck. By putting these textured metal planks under a vehicle's wheels, a driver can get enough traction to climb out of a hole. The ones I bought had been cut from mobile landing strips used by the Allies during World War II. If they served me as well as they had Field Marshal Montgomery, I'd be in good shape.

At Reggane, the last town before the featureless desert where the road ends, the Algerian authorities have set up a checkpoint to keep tabs on who goes into the wasteland and when. If anybody fails to show up at the Algerian–Malian border post of Bordj-Mokhtar, other travelers are warned to keep an eye out.

"We don't think we can let you proceed," said the guard, in impeccable French. "Fitted up the way you are, it seems most likely that you would become stranded."

I had to go on, I told him. The saplings would die if they were not planted soon, and if they died, so would a starving village.

The guard shook his head.

"I will assume full responsibility," I said. "Don't even think about searching for me if I don't arrive on the other side. I'll look out for myself."

After a quick check to make sure I had enough fuel, water, and food, a compass, maps, binoculars, and a fire exinguisher, the border guards let me proceed.

"I bet you a hundred dollars you won't make it to Bourem." The speaker was a tall, skinny German who was cooking his dinner near the checkpoint.

"What makes you such an expert on desert travel?"

"This is my seventeenth trip down there, that's what."

At his side a fat, dark-haired dwarf was fanning the embers of a campfire, every now and then adding a few more twigs to keep the flame from dying out.

"Have a bite with us," said the German. "Drink some tea. And then, if you know what's good for you, go back home with your impossible rig." When he spoke, his tremendous Adam's apple bobbed up and down.

Over soup, many cups of tea, and some dates from Mr. Kheredine's stock, the tall man told me of his business. He bought old trucks in Germany, mostly army surplus, and drove them down to Mali or Niger for a lucrative resale.

"Sometimes I don't even make it with these sturdy monsters," he said, "and they generally aren't even loaded. This year I am taking some chances, as you can see."

His rig and that of his dwarfish friend each carried a smaller truck on its flatbed. In the bed of one of these smaller trucks was a car.

"You have some nerve telling me I won't make it. You're a lot heavier than I am," I said.

"Sure, I'm heavier," he replied, "but the trucks are much higher off the ground than your trailer. When I sink into the sand I might pull through, but your trailer will just sit on its axles. The wheels will simply disappear into the sand. Once I brought a trailer down here, but I had to abandon it in the middle of nowhere just to get through myself."

At this point I had little choice, but I figured if things got too hairy I could ditch the trailer and proceed with the Land-Rover alone.

"Good luck, buddy," the German called out as I left. "See you somewhere along the way. When you break down, you can always hitch a ride with us."

Night had fallen when I left the checkpoint, but the Tanez-rouft Trail has solar-powered beacons every ten kilometers. It was easier to drive after dark, in fact, because I only had to follow the stark glimmers of light. There were long stretches of utter blackness between beacons, but if I kept driving steadily south, I always picked up the next flicker soon.

In two days I traveled 600 miles through the desert, and loved it. Often I'd take my hands off the steering wheel and drive with only my foot on the gas pedal, since the pull of the trailer kept the Land-Rover going virtually straight. Only once did I get stuck, and then I was able to pull myself out fairly easily with my sea anchor and winch.

I ate dates, drank water, sang songs, and made speeches: I addressed a full session of the United Nations about the feasibility of a trans-Saharan railway. I played devil's advocate as well, attacking the plan as an impractical waste of money, but countered each of my own objections with ease. The assembled delegates rose to give me a thundering ovation. Later, I couldn't remember a single one of my eloquent arguments.

The first person I met on crossing the Malian border was a small boy with a bottle of beer. I was driving slowly as I entered the small cluster of mud huts that made up the town of Tessalit; the kid ran up when he heard my engine, to make sure I spent my money at the café run by his father rather than at a rival establishment. After several weeks in Islamic Algeria, the beer was like heaven. Most Malians are Muslims as well, but they are not quite as strict as their neighbors to the north.

But I had barely taken a gulp of my beer when a little man in a ragged uniform burst in and started shouting a stream of abuse at me in French:

"Shit of a dog! You think you *own* this country? You think you can enter and leave as you please? We don't need damn dogshit tourists here! I'm taking you down to the station this instant!"

Usually I'm the soul of politeness when dealing with irrational officials—the only way to get rid of them is to smile and nod pleasantly. But after several dry weeks, anybody trying to separate me from my beer was in for a fight.

"The only dog-shit I smell here is YOU!" I yelled back at him in fluent French. We flung insults at each other until we both were hoarse. The argument ended when the little bastard reached for his pistol but found the holster empty. He shrugged and went to find his superior.

The commanding officer of the Tessalit garrison was a big Tuareg. He made me unload my trailer, and looked over my saplings suspiciously.

"What is all this supposed to be?" he asked.

"Trees for the village of Araouane."

"What organization are you working for?"

"None, I am doing this on my own."

"Why are you doing this?"

"Because the villagers are destitute."

He took me into his office and poured me some tea.

"I do not think you can manage the trail to Bourem with that trailer. But with a guide you might be able to get through the Tilemsi Valley, which is longer but much easier. Guides are expensive. So many Europeans come here thinking they will travel solo but decide they need assistance at the last moment, so experienced guides can name their price. If you cannot afford this, perhaps you could wait for the next army convoy."

I told him the trees could not wait, and he sent a man to fetch a guide.

In the meantime, the commander ordered Officer Dogshit and a few other policemen to help me repack my plants. We wiped the coating of dust off their leaves, and bundled them as securely as possible for the rest of the journey. These border guards became my first volunteers.

While we were busy with our preparations, two familiar trucks rolled in.

"I'll never make it with the trailer, eh?" I taunted.

"I said Bourem," the skinny German replied. "Up to here it is easy."

When my somber black-robed guide arrived, I left the German and his friend to the mercy of customs.

During the three-day drive to Timbuktu, my Tuareg pathfinder barely spoke a word. He merely stretched out his hand to show me the direction to take. We plowed through interminable stretches of sand, the rig sloshing along like a sled in wet snow. Often we barely made ten miles an hour. When we got stuck in one sandpit that seemed bottomless, my sea anchor again saved the day. Its 60-foot cable let us extricate ourselves ten times faster than sand ladders would have, and for the only time in the journey I saw my companion smile.

The Tuareg did not approve of my habit of driving seventeen hours a day. Like most men of the desert, he could not go long without his tea. Every few hours he wanted to stop and brew a pot, while I wanted to make the trip with as few interruptions as possible. The saplings were constantly on my mind, something the Tuareg simply could not understand. It must have struck him as ridiculous for someone to inconvenience himself for a few miserable plants.

When he realized that I could not be persuaded to stop just for tea, he asked in a way that I could not refuse. He became a devout Muslim. Each time he wanted me to stop he would say, *"Prière."* The Islamic faith requires prayer five times a day, so I could hardly refuse. But I strongly suspect that he had never worshipped so many times in his life as he did in those three days. At each break he'd light a fire, set the teakettle on it, and kneel down in the sand just long enough to let the water boil.

We made it safely to Timbuktu, but never saw the skinny German at the other side. Somewhere out there in the desert (I told myself) is a trucker who owes me $100.

TRIBULATIONS

The first person I went to see in Timbuktu (after my obligatory visit to the Welcome Wagon, of course) was Salah
Baba. My laconic Tuareg companion, who must have been
exhausted from his prayer schedules, had bid me farewell at
the first sight of a teakettle in town.

Salah Baba had offered to come back to Araouane with me
and explain the project to the villagers. He was happy to help
his hometown, he said, but I couldn't help suspecting his
motives.

"You will need to bring quite a lot of food," he said. "People cannot work well when they are hungry, and as you
know, there is virtually nothing left to eat there. Furthermore,
you will have to import all the laborers. For the next five
months any Araouane men of working age will be at the salt
mines." From September to March, he explained, all the able-
bodied black men of the village worked in the mines. During
the hot season they generally returned home. I would come
to understand the truly desperate circumstances of their employment later, but it was clear from the look of the town
that they weren't exactly thriving.

He presented me with a mason named Garba, whose daily
wage would be $3 a day plus food rations. Soon he'd also
lined up a mason's aide at $2 daily and four hired hands at a
buck and a half apiece. Garba spoke a bit of French, which
would be very useful once Salah Baba went home. He
beamed at me and smiled a toothless smile. Obviously he
thought he was receiving a royal wage.

"I will do anything you ask of me," he said, "and Salah Baba will punish me severely if you are not satisfied."

Next we went out to purchase food. The prices Salah Baba quoted me meant nothing at all, since I had no basis for comparison. I had to trust him, although since his "assistance" with the caravan, I did not find him terribly trustworthy.

We bought huge sacks of rice, millet, and cornmeal, industrial-size cans of oil, sardines, corned beef, 50-kilogram bags of sugar, tea, and baobab powder. We also stocked up on straw mats, ropes, and wooden sticks to make enclosures for the saplings, to save them from being eaten by camels.

Soon word got around town that a *toubab* was on a spending spree. A mob of vendors appeared out of nowhere, and followed my every step trying to hawk their wares. One con man offered to sell me several crates of dehydrated minestrone soup "below purchase cost." When I examined the merchandise, I saw that each packet bore an inscription in several languages, warning that the food was a gift of the Italian government to the school system of Mali, and was not to be sold or exchanged. I threatened to go to the police, but neither the vendor nor anybody else seemed the least bit disturbed.

By the end of the afternoon, I owned 7 metric tons of supplies, the maximum load that a desert truck could bear, I was told. When I drove to the warehouses to load up my supplies, I found that almost everything I'd purchased through Salah Baba had been filched from international charities. The rice and cornmeal bore the legend "Donation of the United States," the sardines were from Japan, the oil from Finland, and the corned beef from Denmark. Do not worry, all the bystanders told me, that is how it is here. Nearly everything for sale at the Timbuktu market was stolen from relief agencies.

What could I do? I'd already bought the food, and no amount of arguing would get it to the tables of the hungry children where it belonged. By bringing it to Araouane I'd at least be able to feed a few hundred people who would otherwise have gone hungry themselves.

I hated the idea of working with Salah Baba even more now, but I could not manage without his help. And despite his profiteering, he seemed a little less of a cheat than anyone else I'd met in Timbuktu.

With the merchant at my side, I hired a truck, a driver, a guide, and three assistants to help dig the vehicle out when it got stuck in the sand. More money for gas, oil, transmission fluid, and the crew's food and lodging both ways—I just kept my wallet open and paid what Salah Baba told me to pay.

We showed up for the trip out at dawn the next day. The driver was waiting for us in the middle of the biggest, stinkiest garbage dump in all of Timbuktu. I was shocked to find the truck loaded not only with my dearly bought provisions, but with at least twenty passengers as well. I'd been told that the vehicle could bear no more than 7 tons, but this mob and their voluminous luggage must surely have doubled our weight. They were all salt miners whom the driver had charged a lot of money for transportation to Araouane, from which point they'd continue on by caravan to Taoudenni. They hadn't known about me, and I hadn't known about them, but at this point there was little for any of us to do.

Salah Baba and I drove alongside in my Land-Rover. Overloaded as it was, the truck made excruciatingly slow progress. At this rate, I said to myself, we won't reach Araouane before the end of the month. I asked Salah Baba if he could find his ancestral village without the guide.

"Eventually I'll probably find it," he replied uncertainly. His pride wouldn't have let him say otherwise.

We followed a trail of camel turds, on the assumption that if loaded camels could get through without being sucked into a sand hole, so might we. Sure enough, after a few kilometers we reached harder sand, where recent tire tracks had not yet been wiped away by the wind. Île de Paix, an aid organization, was building a well north of Araouane, and the tracks were from their supply vehicle. It looked as if we would reach Araouane by nightfall.

We lost the tracks around noon, however, and the Rover got stuck. It was in too deep for the winch alone, so we had to dig our way out slowly, painfully, 6 feet at a time. I did not particularly like sweating away under the hot sun, but I did find comfort in watching Salah Baba do the same. The marabout had always relied on servants for every physical need, and I guessed from his soft, almost feminine hands that he had probably never held a shovel or any other tool in his life.

To add insult to his injury, we had no tea. Salah Baba had spent much of the morning grumbling about what a grave oversight it had been to set out without a kettle, and now he started showing severe symptoms of withdrawal. He muttered to himself, panted and huffed, smoked the last of his cigarettes, and glumly sipped warm water from his goatskin. We worked until dusk, then had dates for dinner.

A sickle moon, like the prow of a Venetian gondola, illuminated our camp and threw shadows on the long trench we had spent the afternoon digging. "Another hour or two of shoveling tomorrow," I said, "and we'll be on our way again." Salah Baba just nodded miserably.

We were too tired (and I was too excited about our arrival in Araouane) to hope for much sleep. Salah Baba and I chatted far into the night. I fashioned some cigarettes for him by gathering up bits of tobacco from the car's glove compartment and wrapping them in scraps of toilet paper. Indirectly, he admitted that most of the Araouanite merchants I'd met in Timbuktu had been living for years off donations that were supposed to be sent to the village. When applying for charitable contributions, they'd list Araouane's population at 3,500, knowing full well that nobody would bother to visit the remote settlement and check. When rations arrived in Timbuktu, they'd sell what they could in the market, pocket the proceeds, and send on supplies only to their own Moorish kinsmen. They would have liked to feed the whole town, he protested unconvincingly, but regrettably they did not have enough money for transport.

He told me stories of past travelers. A caravan with 200 men and 2,000 camels had followed the trail down from northern Tindouf in search of Araouane's fabled wells. Somehow they strayed off course, wandering blindly for weeks. Every man and every camel perished only a few kilometers from the village. This had happened when he'd lived in Araouane as a little boy, and the memory of it had left him a bit afraid of the desert.

In the morning, the soft dunes of last night had all been rearranged by the wind. We finished digging ourselves out, but had no idea which way to go. We had a compass, but how to set a course without knowing the starting point? We could travel in a straight line, but we did not know which straight line would take us to Araouane. Salah had no helpful advice, so the only thing to do was to make an educated guess: I knew that Araouane should be about 260 kilometers from Timbuktu, and supposedly the driver of the truck whose tracks we'd been following had been taking a fairly direct route; since we had pretty much continued on his course, we'd just keep driving in the same direction. If there was nothing in sight when the odometer read 260, we could begin to worry.

By early afternoon we began to worry. The dial stood at 263, and Araouane was nowhere to be seen. Of course, the sand plumes made visibility so bad that we could have been on top of the town without catching sight of it. I suggested that we make camp until the wind died down, and then drive around in wide circles. Salah Baba could not wait. We had enough water for many days, and enough food for many weeks, but he couldn't survive another night without tea or cigarettes.

"Let's drive around some more," he said. "You never know, we might get lucky."

I didn't want to use up too much diesel, just in case we did not find Araouane and needed to drive back to the Niger River. But just to keep Salah from having a fit I drove on, trying to pierce the swirling sands with my weary eyes.

After an hour or so, Salah gave a gleeful shout. "Over there," he cried. "People!"

I saw them, too, vague phantomlike shapes on the horizon. They turned out to be two men and a little boy taking a few camels from their encampment to the wells. Salah was blubbering with joy, and I probably would have been too, if I'd been able to blubber in a language they could understand.

The boy came with us, to lead us to town. We felt foolish having to make our grand entrance under the guidance of a small child, but at this point we had little pride left. After about six kilometers we saw the first house, the very same one that had briefly been my home.

"*Hamdulilla,*" Salah muttered.

"Thank God," I echoed.

The population must have heard the car's engine, for by the time we arrived, almost the whole village had assembled on the nearest dune. They didn't quite know what to make of my strange rig, but when they recognized Salah Baba, they launched into the standard interminable round of greetings:

"Peace be with you."

"Peace be with you also."

"May God protect you."

"May God be with you."

"No evil to you."

"None to you either."

"I trust your health is in the hands of God."

"I hope for yours also."

"I trust you are not tired."

"God has provided us with a safe journey."

"God be praised."

"May your sons be in the hands of God."

"May your sons be in the hands of God as well . . ."

Because of Salah's noble ancestry, the salutations went on even longer than usual.

Midway through the formalities Salah started gasping for

tea, and some children ran off to fetch a pot, sugar, and cups. He downed glassful after glassful, barely slowing his words to drink. After all, he had two days of catching up to do.

Salah talked and talked, and the villagers mostly just listened. I didn't understand a word, of course, but afterward the marabout told me what he'd said. "I explained it all to them," he drawled through his tea. "I said that they had but two choices, to stay here and follow your instructions, or to go to Timbuktu and join the horde of refugees there living off handouts. I told them that many of the things you would tell them to do might seem strange, but that you came from a strange country where everybody had enough to eat. I told them that if they all worked hard and did as you said, they too would have more food than they could ever imagine."

"And how did they respond?"

"They did not say much. But they know it is their last chance."

With Salah Baba giving the orders, things went very smoothly. We unloaded the car, then drove the trailer down to the wells to get the saplings watered and cleaned. But the villagers had absolutely no concept of order. Try as they would to treat the supplies with care, they quickly turned all the area around the storehouse into complete chaos. Cases, boxes, and bags lay about helter-skelter, as if the stuff had been dropped from an airplane. Somebody deposited my toolbox upside-down, sending a flurry of nuts, bolts, nails, and miscellaneous odds and ends clattering into the sand.

We gave all the plants a thorough cleaning. One ragged little boy dogged my steps, quietly offering a goatskin of water whenever I glanced in his direction. The townspeople were mystified by the saplings, because they had never in their lives seen a young tree. Apart from the aged camel thorn by the mosque, most hadn't seen any tree at all. It was difficult for me to explain how these apparently lifeless twigs would one day yield food.

I demonstrated how to rinse each green stalk gently, make

sure all the roots were protected, and pack the soil back into the traveling cans. The materials for building the enclosures and seedbeds were all on the truck, so we wouldn't be able to start work in earnest for a few days.

Salah Baba had wanted to visit the tomb of his ancestors, but he fell into a deep sleep even before evening prayer. The villagers had brought us their nicest blankets and spread them over the sandy floor for us.

The boy in the threadbare rags, the one who had followed me around with the goatskin, remained huddled by the door all night. "Doesn't he have to go to his home?" I mumbled before dropping off to sleep. But Salah was already snoring.

When I woke up in the morning, that same little boy was still sitting by the door. He had lighted a camel-dung fire and put a kettle on for tea. Two buckets of clean water for washing up, which he'd lugged all the way from the well without disturbing us, stood at the foot of our blankets.

The boy never spoke and seldom smiled, but his eyes displayed a lively intelligence. He watched my every movement eagerly, to see how he could be of use. As soon as I started tidying up my heap of boxes and bundles, he joined right in, somehow managing to lend a hand while still busying himself fixing our breakfast.

"Who is this boy?" I asked Salah Baba.

"His name is Bou-djema," the marabout replied, pronouncing it like "bushman." "He has never seen his father—the man was a miner at Taoudenni, who made the boy while passing through town; he probably does not even know this child exists. Bou-djema's mother is busy caring for two little daughters by other men. The boy is still too young for the salt mines, so he helps with visiting caravans and tries to get food for his grandmother."

Bou-djema was the first Araouanite I came to know by name. Looking back on it all now, I think that the entire garden project might have failed if not for his unflagging work.

• • •

A few days later, while I was laying out the garden fence with Bou-djema, a man came walking out of the dunes from the south. He said that our supply truck had run out of diesel about fifteen kilometers from Araouane.

The driver apparently had bought less fuel than I'd paid for, and pocketed the extra money. I hated to give up any of the diesel in my own car, since I'd need plenty for the trip back to Timbuktu, but there was little choice: all our provisions were stuck in the desert, and I was paying for the truck and crew by the day. I loaded the Land-Rover with a few jerricans of diesel and drove back to the truck. When the driver asked for more money, saying that the trip was much more expensive than he had thought, I flew into a furious rage. I imagined how he had mused on the trip up how he could further milk the cash cow—me. I yelled:

"What hotel did you shitty bastards stay at, the fucking Sahara Hilton?" I screamed, veins popping out of my neck. "Ask me for one more fucking franc, and I'll pound in your fucking face!"

So much for winning my coworkers' affection. But my outburst did make its point: When the time came for the truck driver to go back home, he must have miraculously discovered some extra fuel in his rig, because he didn't ask for any more for the return trip.

My next task was to distribute the food to the people. I had no intention of simply doling out rations to every starving person I found—Africa was too big to be fed by me. I would give food only to those who were willing to work, but first they needed a chance to build up their strength. I told everyone who intended to help in the garden to show up at my house for their signing-on bonus.

Pouring the millet, cornmeal, and baobab powder made such a dust cloud in my room that I could barely see. About forty-five people claimed rations, shyly whispering thanks in Sorai, barely daring to look at me as they held out their cups and pans. Bou-djema, of course, was there to help me, his face soon a ghostly white mask with big, red-rimmed eyes. He measured out each portion with earnest concentration.

At seven the next morning we got down to work. Garba, the mason, went into the desert with some of the village elders to look for building materials. The Araouanites showed him where to find underground deposits of banco, the dry claylike mud with which we could make bricks to form a wall. Some of the men and children started to dig holes for the posts, but the women just stood around watching.

"Why don't they help?" I asked Garba.

"As soon as we need to have somebody carry water, they will work," he said. "That is all the women are permitted to do."

The kids were the most diligent workers by far. They watched me plant saplings and tried their best to imitate my actions. The adults, on the other hand, generally did as little work as they could possibly get away with. The rations I had distributed included tea and sugar, and most of the men could not resist the temptation to sit and drink.

It was a sad spectacle—full-grown adults taking their ease while the children labored in the hot sun. When I had Garba admonish them for me, a big, strong man named Fah shouted back that nobody could tell him when he could or couldn't have his tea.

"Go off and drink, if you like," I told him. "But don't come back. I hope you like eating lizards, because you are not going to drink my tea anymore."

The next day he apologized and asked to be reinstated in the project. I turned him down—Fah was too lazy to do the work of even a small child, so I decided to make an example of him. The project would not be a handout, I was determined.

After that, the men spent more time tending the plants and less pouring the tea.

We got all the saplings planted, but the sudden appearance of our little patch of greenery in the midst of the vast wasteland attracted some unwelcome visitors. Locusts arrived in swarms. The villagers ate them at first, but now that they had rations to fall back upon, they began to see the insects as a nuisance rather than a staple. I could see why: I fried up a few and forced myself to eat them, but they tasted like shrimp that had gone bad. Only one old woman actually liked the creatures; even much later, when we had chickens, goats, camels, and all sorts of vegetables in abundance, she stuck faithfully to a diet of locusts.

In retrospect, our initial progress seems incredible. With the exception of the six hired men from Timbuktu, none of these people had done any gardening work before in their lives; most of them had not done much work of any sort, because there had been precious little to do. Yet within a few short weeks, they had planted all the trees and were keeping them regularly watered and fertilized. They had built crude protective fences around the stalks with the material I'd brought from Timbuktu, so that the hungry camels and goats could not eat them, then sturdier walls of banco brick around the whole to provide protection from the violent desert storms that would have sand-blasted the plants to mulch.

But I gradually realized that I would not be able to leave Araouane any time in the near future. The villagers were willing to do virtually everything I asked of them, but they hadn't yet acquired either the initiative or the know-how to sustain the project on their own. And I still sensed that they were going through the routine of farming merely to humor me. To them, my whole project remained ridiculous.

The fact that I couldn't express myself to them directly didn't help. Every time I wanted to say something I had to send Bou-djema (who was never far from my side) off to fetch Garba. Even then communication was difficult, because the mason's French was sketchy at best. It took so much effort to arrange to get my message across that I often forgot what it was I'd intended to say.

When I tried to explain through Garba that the fragile

saplings of fruit trees would one day produce real food, the villagers looked at me with mild amusement. When I had them poke holes in the ground, drop tiny seeds in, and pour water on top of them day after day, I no doubt confirmed their impression that I was a lunatic. But they were good-natured enough to follow my directions anyway.

"What the hell," I imagined them saying to one another, "so long as he's providing food."

When I made watering cans by poking holes in some empty sardine tins with a nail, however, I'd evidently gone too far. The old black village marabout, Baba Cambouse, took me aside with a delegation of elders.

"We know that you are here to help us," he said through Garba, "and we are very happy about this, and we thank you very much. But maybe some of the things you do are not very good. These cans can be used for storage, but now you have destroyed them."

Garba could barely keep himself from laughing, since Timbuktu is practically blanketed in garbage-heaps of rusting tins. But there was nothing funny about our total lack of communication. I simply had to make the people of Araouane understand just what the project meant and how to keep it going once I returned to America. For that, we would have to speak the same language.

I decided to teach the local children some French. After all, they'd need French to communicate with any outsider or government official. I had Garba summon all the kids to my house in the evening, promising dates and peanuts to whoever showed up.

They started arriving while I was preparing my dinner. Soon twenty pairs of eyes were following every move I made. As I ate, they stared in wonder at my little gas cooker. How could there be fire without camel dung?

We went outside to a nearby dune and sat in a circle. I pointed to myself, and said in a slow voice:

"*Mon nom est Aebi.*"

A look of confusion ran across their faces.

"*Mon nom,*" I repeated even more slowly, gesturing wildly at my chest, "*est Aebi.*"

The looks of confusion only deepened.

I suddenly remembered that *abbi* is Arabic for "my father." I was, in effect, telling them that I was their daddy.

I tried a new tack. I had Bou-djema stand up, and pointed at him.

"*Toi,*" I said, "*tu es Bou-djema.*"

"*Toi,*" he repeated dutifully, "*tu es Bou-djema.*"

"*Non, non. Je suis Aebi. Toi, tu es Bou-djema.*"

"*Tu es Bou-djema.*"

We had a long way to go.

I had the children come every afternoon once work was done. One day I brought out a clock and showed it to them. When I made the alarm go off, the room was full of smiles. They loved the chimes, but had no idea what it all meant. How, I asked myself, did you explain a clock to somebody who has no concept of hours and minutes?

I hit on the idea of prayer times. Observant Muslims are supposed to pray at five specific times each day. The Araouanites might reckon these hours by the sun, but they reckoned them all the same. And since most religious concepts retained their Arabic names, I had some common ground for communication.

"*Salat el sobbah,*" I said, and made motions of praying. That they understood. I set the clock's hands to sunrise, six o'clock. "*Il est six heures,*" I said, and had them repeat it.

"*Salat el soehor,*" I explained, and I set the time for noon. "*Il est midi . . .*"

I wasn't sure how much they really understood, but the alarm never failed to get a laugh.

Occasionally old Baba Cambouse would drop in on the daily lesson. And as the children started to speak a few French phrases here or there, so did Baba Cambouse. Gradually, though, I realized that some of the words he used were ones I hadn't taught the kids. Within a month, the old man was speaking almost fluent French loaded with argot.

Baba Cambouse was the town's oldest resident. None of the villagers knew their exact ages. They were simply born "in the year when the locust invasion was terrible," or "in the year when many camels had died." But Baba claimed to be about eighty years old. He still stood straight as a ramrod, and despite the camel turds he kept stuck in his nostrils to combat a chronic drippy nose, his mind was sharp as a tack. Finally I asked him how he came to know the language.

He told me that when he was a kid—"*un gosse,*" he said, using the colonial slang—a garrison of the French Foreign Legion had been quartered in the very house which I now occupied. Baba had earned bits of food by doing odd jobs for the soldiers, and picked up enough French to understand their requests. He hadn't spoken a word of the language in nearly seventy years, but now, he said with a toothless grin, it was all flowing back to him.

Bou-djema continued to prove himself invaluable. He rounded up all the kids before each lesson and got them in line without yelling and screaming. When I finished distributing the school snacks each day, he took the remaining dates and peanuts and, without a word of instruction from me, returned them to their storage sacks.

Each day when I came back home, I found two buckets of clean water outside my door. I barely finished my evening meal before he had washed out my soup bowl. By the time I'd turned around, he had folded my blankets and tidied my bedding. When the other children left after their lesson, Boudj (as I started to call him for short) simply sat waiting until he could offer some help. When I went to bed at night, he was sitting quietly by the door, and when I woke up in the morning, he was again sitting there waiting to be of use. Did he ever sleep?

I liked the boy very much, but after a short time in Africa I, like most foreigners, had grown a bit leery of people's motives. I decided to put the boy to a few tests. I "forgot" a knife under a bag, "misplaced" a few dates in a dish, a T-shirt in my rubbish heap, a bag of sugar outside the house. Boudj pointed every item out to me.

Yes, I said to myself, this kid was going to be my right-hand man.

The lessons continued, and I tried hard, but I knew that I was just not suited to be a teacher. I didn't have the training, and, more important, I didn't have the patience. I put out the word in Timbuktu that Araouane was looking for a school-master.

At first it seemed unlikely that we'd be able to entice anybody to the village. After all, most of the region considered Araouane a starving hellhole engulfed by sandstorms and swarms of flies. But on one of my monthly trips to Timbuktu to haggle for supplies with smugglers, thieves, and cheats, a young man approached me in the street.

"I hear that you are looking for a teacher," he said.

"Yes, I am."

"I am a licensed instructor from the Malian Teachers College," he said.

"What are you doing now?"

"I am unemployed. The government is not hiring any teachers, and not even paying those already on the payroll."

"So how do you spend your time now?" I assumed that he was one of the hordes of idle men who sit around in the alleys all day, drinking tea and gossiping—very slowly, of course, so as not to wear themselves out.

"I do athletics," he replied. "I am the regional eight-hundred-meter running champion, and helped put together the Timbuktu soccer team."

"What is your name?"

"Mohammed Ali Ould Ahmed."

He was a tall, intelligent-looking Moor with a very light complexion. Knowing the local whites' racist attitudes, I warned him that he'd have to be living in close quarters and on equal terms with a population that was mostly black. The news did not bother him at all. His whole childhood, he told me, had been spent in a little black community outside Timbuktu. His family had been one of the poorest in the village, and the only Moors for miles around. At his college in Sikasso almost all his friends had been black. He felt closer to them, he said, than to people of his own race.

I offered him about $160 for one month's work, with the understanding that if I wasn't fully satisfied, I'd send him right home. He didn't ask for an advance to give to his ailing mother or a food ration for his starving children or a loan to pay for teaching supplies.

"I'll be on the next truck," was all he said.

It is hard to imagine Araouane without Mohammed Ali. Within two years he taught the children to speak French nearly fluently, and gave them a knowledge of local and world geography as well. He taught them hygiene and set up athletic programs. Eventually he began to teach French to the adults, and to show the women and children how to prepare the strange vegetables we were growing. With the help of Bou-djema, he kept track of all food rations and supplies. During his second year's vacation, he studied first aid at the Timbuktu hospital so that he could treat injuries and dispense basic medical treatment to the villagers and local nomads.

By the third year, Mohammed Ali had taken charge of a general store we set up to sell matches, turbans, cloth, candles, sugar, tea, tanning acid, soap, and other goods to nomads from all over the desert. He made enough profit from the merchandise that I didn't have to pay him a salary anymore, but despite his monopoly, he never seemed to have much money. He was always helping friends out with "loans" he didn't expect them to repay. He let the nomads pay off their bills with sheep, goats, or camels if they chose, and instead of eating the beasts at the first opportunity, as had been the local custom, he built up a nice herd that gave the town a ready source of milk, cheese, and meat in times of scarcity.

By the end of the first year, the garden had become meaningful to some of the villagers. They had found that the little specks they had buried in the ground really produced food—strange food, but food just the same. At the end of three years, Mohammed Ali had practically put me on Araouane's unemployment list.

SLAVES FOR SALT

Around the beginning of May, I went back to New York, leaving the project in the hands of Mohammed Ali. All through the summer, while I lounged in air-conditioned comfort, I worried about how my new home was holding up.

I did not have much apprehension about the garden itself: I'd taken the mason, the mason's aide, and the four laborers back to Timbuktu, but Mohammed Ali had earned my full trust and by now some of the villagers were almost sold on the project. My main concern was relations between the Arabs and the blacks.

Throughout the Middle Ages, the whites of the Sahara had enslaved the blacks of regions to the south, either to keep them for their own use or to sell at great profit. Sometimes the nomads came close to treating their slaves like members of the family. Often, though, the servitude was far more onerous. The eighteenth-century explorer Mungo Park left this description:

> *The slaves in Africa, I suppose, are nearly in the proportion of three to one to the free men. They claim no reward for their services except food and clothing, and are treated with kindness or severity, according to the good or bad disposition of their masters. . . . The domestic slaves, or such as are born to a man's house, are treated with more lenity than those which are purchased with money.*

Park reported that "bought" slaves could be sold to strangers at will, while those born in a master's house were never alienated from the family. A slave's value increased with the distance from his native town, on the belief that the farther away he was from home, the less likely he'd be to attempt an escape.

For centuries the Arabs of Araouane had literally owned the village and all its inhabitants. Even at the time of my arrival, most of the black villagers had lived their lives as bond servants. Though slavery had been officially abolished in Mali since French colonial times, the remote towns and hamlets operate virtually as they did in the days when Timbuktu was a center of the African slave trade. The ancestors of the blacks had been brought here hundreds of years ago to wait upon the Arab caravan owners, and when I arrived, I found them still doing their masters' bidding.

Technically this slavery is self-imposed. But because there is no work for a man in his village, virtually his only option is to sign on with an Arab "patron," who will pay him to labor in the salt mines. But the patron charges his worker for transportation to and from the mines, for food and clothing, and since he has a complete monopoly over these things, he soon puts the worker in a state of permanent dependency. In order to supplement his starvation rations, the miner must buy extra food on credit from his patron, which sinks him deeper and deeper in debt. The longer he works, the more indebted he becomes.

Each miner traditionally works six to seven months of the year at the mines, for that is the longest time a young man can labor there without suffering irreversible damage to his system from the poor quality of the available water. When he gets too sick to work, or too old to go to the mines, his children inherit the debt and have to toil for the patron their whole lives to try to pay it off.

Mohammed Ali explained to me that if he were to have a child with one of the black women in Araouane without having gotten the consent of her master first, the child would belong to that master and Mohammed Ali would have no claim on it. If it was a boy, he would grow up to slave away in the mines like all the other black men.

Around the middle of March, the miners had started to come back to the village, a few of them tagging along with each caravan that passed through. I was shocked to see their condition. They had walked the 500 kilometers from the mines, since all the camels were fully loaded with salt. Most of them could not afford shoes, so they'd made the voyage as they'd labored for months, barefoot. Their feet were covered with cracks and gashes so deep I could see bone. One man begged me for medicine to soothe his feet, which he'd sewn up with strands of nylon from an old flour sack.

Before I left for the summer I'd told Mohammed Ali to observe which of the miners were the most willing workers, which ones had the biggest stake in the village, which took the most pride in their families and their environment. If I was going to help Araouane become independent of the outside world, I would have to help find a way to keep her men in the village.

When I returned in the fall, I brought with me Fritz Gross, an artist friend who had put aside his painting and sculpting to help with my project. Mohammed Ali had come up with a list of six local men to replace the six hired laborers we had brought in the previous year and sent home with the hot weather. Their patrons were due from Timbuktu shortly, to escort the men back to the mines, where they'd take turns supervising them.

Before the masters arrived, Fritz, Mohammed Ali, and I took the slaves aside and made them a proposal: We offered to pay off all their depts and free them from bondage if they would stay in the village and work on the project. In exchange they would get ample food rations, medical care, clothing, and education for their children, as well as for themselves and their wives if they so wished. At the end, in exchange for his contribution, every man would receive as

his own property a portion of the communal garden, a house
on the perimeter of the garden, and a stake in the income
from the eventual tourist trade.

We might as well have spoken in Chinese. The men had no
concept of income or ownership. From their point of view,
they were only being "sold" from one patron to another.
They thought that they were being assigned a plot of land to
tend—for *my* profit. Trying to explain "income from antici-
pated tourism" was a complete exercise in futility.

And why, they asked, with admirable prudence, should
they desert masters who at least provide them with work and
food in favor of some foreigner who might abandon them at
any moment?

Mohammed Ali thought that he could get the idea across
to them better. He explained that their master would not be
me, but the village, of which they were a part. He told them
that half of their freedom price would be paid by Fritz and
me as a gift, half would be owed by them to the village;
moreover, they would only have to pay back the village's half
if our predictions about the economic viability of the garden
proved true. That seemed fairly straightforward—except that
the local concept of money was salt bars, and none of these
men had ever owned any other type of currency.

"How are we to pay Araouane," they asked, "if we cannot
cut salt?"

Eventually they graciously permitted us to purchase their
liberty. But I think that they only gave consent because a
mere six of them had been singled out. They had little concept of freedom as yet, but they were proud to be among the
elite chosen to receive even an inscrutable reward.

We invited the patrons to dinner. But I had decided that to
keep the merchants from banding together to gouge me on
the prices, I would have to gain the upper hand right from
the start. Back where I came from, I would have tried to set
potential business partners at ease, by creating as pleasant an
atmosphere as I could. I insisted that we meet at my house,
offering them transportation in my 4x4—for the fee of one

goat per passenger or 100 kilograms of luggage. The goats would later be distributed to the workers in the garden.

I ushered the patrons into my house. By this time I'd made my digs very comfortable, with solar-powered lights, a table, a solar-powered fan, and the tape deck from the Land-Rover, which was playing the Arab equivalent of easy listening. Mohammed Ali explained our plan to them, and five pairs of greedy eyes lighted up with glee. A deal to be made!

Then they heard my terms, and their smiles faded. They claimed—rightly—that we were trying to get their best workers. We countered that soon there would be no market at all for salt and then none of the labor would be worth anything.

"There will always be a market for our salt," they replied.

They were obviously not going to concede much, so I cut in before they could name a price.

"You all claim that the survival of your village is foremost in your hearts," I said. "What we are trying to do is make certain that your ancestral home doesn't get wiped off the map. It all costs you nothing. I am not asking you to give us your men, I'll pay for them. All I am asking is that you consider the alternatives if you do not cooperate."

They looked at each other a bit nervously.

"In the future," I continued, "you too will be able to benefit from the Araouane we have in mind. If it becomes a thriving community, you, the original leaders, will surely get great prestige and profit. But if you establish yourselves as enemies of its progress—and progress we shall have, with or without you—that will be remembered. It's up to you gentlemen to decide where you want to stand."

I had each of them write the price of his man on a piece of paper. I told them that I was not going to negotiate, that I would not accept second offers, that I would simply accept or reject. Since we all knew that they could get as many unemployed men in Timbuktu as they wanted and that the price of salt had fallen so low that it was hardly worth cutting and transporting anymore, they had a strong incentive to see the deal go through. All of them knew by now that I kept my word religiously and that if I said I would not accept any second offers, I meant it.

The average price came out to about $450 per man. We accepted all and paid, and the patrons probably made themselves miserable wondering how much more I would have agreed to. I guess they will never know, because I don't know either.

That season in the village, the blacks of Araouane began to see themselves a little differently. They had been so accustomed to accepting their station as the "natural" servants of the Arab patrons that it was hard for them to absorb the notions of equality and autonomy. They needed examples.

The best example came from Mohammed Ali, who (although an Arab) worked harder and longer at everything than anyone else in the village. I too undertook all manner of manual labor—but the townspeople already thought I was touched, so perhaps my digging ditches did not impress them quite as much. Fritz, always a hard worker, combined every activity with jokes, so the adults assumed he was playing, but his method worked wonders with the children.

On one of my trips to Timbuktu, the ever-present street kids, who by now had given up begging *cadeaux* from me, came running up:

"Ernst," they said, "your son has arrived, he is at the police station."

I went right away. Sure enough, there was Tony, crowded into that stinky hole with five young French travelers who had taken a *pirogue* down the Niger River from Mopti. He looked great, considering he'd been cramped in a little boat for almost two months.

The policemen were just about to extort their sum when I claimed my son. By now I had a friend in Lamine Diabira, the governor of the region, who had on various occasions shielded me from bureaucratic assaults. Miraculously, the "fee" was waived and passports returned with forced smiles and terse wishes for a nice stay.

Tony came with me to Araouane, where he spent a long month working like a dog. It was wonderful for the villagers to see a man whiter than the whitest Arab in town sweating under the same heavy loads as they did. Tony figured that since he was there for only a short time, he could really run himself dry to show that hard work was not degrading or dishonorable. Sometimes he worked so hard that in the evening one of the children had to walk around on his back to massage his knotted muscles into shape for labor the next day.

Tony and I had never before spent so much uninterrupted time together. During the long nights without radio, newspapers, friends, movies, restaurants, telephones, or any other diversions, we grew much closer than we ever had been back in America. The village kids all fell in love with him. At the end of each day he would take all those who had worked particularly hard, black and Arab alike, for a drive in the Land-Rover, and sometimes he let them take turns steering.

Come January, though, Tony had to return to school. On New Year's Day, just as we were about to leave for Timbuktu, Bou-djema came running up with a millet sack full of his clothes on his head, dragging a couple of empty goatskins for water.

"Please, Ernst," he said, "lend me a camel to carry the water. I want to go help Tony work in the gardens of America."

In Timbuktu, as I was preparing to drive back to Araouane, an eerily familiar red 4x4 Mercedes pulled up. Before recognition could set in, my brother Peter jumped out of the vehicle. Two months earlier I'd sent him a message requesting netting to protect the garden from locusts, but I had no idea he'd be delivering it in person.

"You must have lost your fucking mind!" were the first words out of his mouth. "You're crazy, stark raving mad! Isn't there a shrink in this town? Are you sick? Gone overboard? Out to lunch?"

"I'm really happy to see you," I said. "You just missed Tony by a few hours, can you believe it? So how was the trip?"

"You're nuts!" he ranted. "Completely crazy!" I looked at him more closely. It was no joke, he was steaming mad.

"Let's get a beer," I said. I left the Rover with the local welder for some minor repairs, warning him to keep a sharp eye on my gear inside. When I picked it up later, nothing more than my binoculars and the toolbox were missing.

With his first cold drink in weeks, Peter's mood improved. He'd driven in a mad dash from Europe to deliver the gear, gotten lost in the desert, run low on diesel, and had traveled nonstop for two and a half days to reach a settlement before his water ran out. In desperation he set off a string of rockets he'd been saving for a New Year's fireworks display for the villagers. An old nomad had led him to Araouane—where he'd been told that I had left for Timbuktu.

"Come on," I said, "take a shower, we'll have a good meal, and when we go back to Araouane tomorrow, you'll really love the place."

"You really are nuts if you think I'm ever going back, and nuts if you want to spend your life rotting in the desert. Why, I should bind and gag you and drag you back home with me, you'd thank me for it later. . . ."

After a long talk and a considerable amount of beer, we ended up with an agreement. Peter consented to come to Araouane for one week, and not a day longer.

Araouane at that time had nine wells left, out of a total of 110. Gradually one after the other had collapsed or been filled in by the shifting sands. In the old days a slave was sent down each water hole periodically to keep it in good repair, but now there were so few caravans coming through that the villagers could make do with only a few functioning wells. It had been so long since they had been maintained that none of the villagers could remember seeing a man go down one.

Peter decided that he was going to renew the tradition. He settled on the well closest to the garden.

Up to this time I had not been aware of the general fear with which the villagers regarded the wells. There is no more frightening prospect to an Araouanite than going beneath the earth's surface, and even the salt miners are terrified of deep holes. At Taoudenni the pits are dug very wide and shallow, since the salt layer is fairly close to the surface. The wells, on the other hand, are no more than 8 feet in diameter at the surface, and much narrower at water level, about 150 feet down.

When I announced that Peter was going down, there was general consternation. The villagers tried to dissuade him with every argument they could think of. Anybody who knows Peter would understand that from then on there was no way he would not go. In his younger days he had been an avid spelunker, exploring many of the most daunting caves in Switzerland, so this puny hole in the sand was not going to intimidate him.

I showed him a few of the collapsed wells, just craters in the sand now, to make sure he understood what he was getting into, and then he started looking for material to build a pulley system so we could let him down. His nonchalance did have limits, though: He drew the line on going local when he saw the ropes used for hauling water, just a few strands of camel skin twisted together.

With all the men and children assembled around the well, we let him down in a millet sack attached to one of the ropes from my stock of gear. For a helmet he wore a couscous basket stuffed with towels. The men who let him down were chattering nervously at the time he was down, but Peter came up none the worse for wear. He reported that the well seemed to be in good shape, but he'd accomplished a good deal more. He'd shown the blacks and whites of Araouane that a white man could do the work of a slave and do it willingly.

Fritz and I had assumed, naively, that the six men whose freedom we had purchased would revel in their newfound independence. It wasn't quite so. When the other men of the village left on the caravan for the salt mines, they watched wistfully and seemed to wish they were going as well. They had made a monumental decision, one that would change their whole lives; they had not yet had time to absorb what it meant. For as long as they could remember, they had simply belonged to this or that Arab family, and in exchange they had been told what to do and had had their basic needs taken care of. Now that familiar arrangement was gone.

Even though Mohammed Ali had repeatedly explained their situation to them, they took up the Timbuktu laborers' practice of addressing me as "patron." Among the villagers it had always just been "Aebi" before. They stopped joking around with me and pouring me the customary three glasses during their tea breaks, as it was not proper for me to drink with them. They no longer came by my house to chat, or brought me little dishes of food specially prepared by their wives. They no longer invited me to their homes.

The Moors' behavior toward me had changed just as dramatically. Although I had tried from the beginning to treat the whites and blacks the same, in their view anyone who socialized with his "inferiors" became one, and they had treated me accordingly. But since our "purchase" of the slaves, Fritz and I had become honorary patrons, part of the ruling class. All of a sudden the Moorish families were competing to see who could have us over for dinner most often.

This competition introduced us to a wide array of fancy dishes, probably far surpassing any that had been served in Araouane for years. We ate young pigeons roasted over *halfa* and then simmered in a piquant sauce with hard-boiled chicken eggs. We had a steamed, fluffy bread called *tacoulah,* dates stewed in peanut sauce, mountains of couscous, and stuffed goat and camel innards. Boudj's grandmother prepared a dish called *kata,* which was something like balls of angel-hair pasta. When word of how much we enjoyed *kata* got around the village, the delicacy was served to us at every meal.

Practically the only time I got to eat with my black coworkers anymore was when I butchered a camel. Tradition dictates that when a camel is eaten, the whole village joins in. Tradition also dictates that the chief gets the animal's heart and other choice organs. Because I was the one paying for the camel and distributing it, I was de facto chief. Camel liver, unlike the livers of most other animals, in my opinion, tastes delicious, and it is always huge. After the butchering, we'd singe the liver over a *halfa* fire and (a twist I added) lightly salt it.

But now whenever one of the former slaves did the butchering, I'd find much of the choice meat missing. Once I caught Zain and Habba literally red-handed, their hands covered with camel blood. When I confronted them with their theft, they were unapologetic: It was accepted that a servant steals from his master, so long as the theft is relatively small. The other townspeople didn't object either, even though it meant their rations would be smaller. When I was just an eccentric outsider, nobody had raided the project's supplies, but as a patron I was fair game.

Our relationship was no longer that of comrades—I was now their boss, I owned them. Whatever they did, they insisted on pointing out, they did for me. They had a desperate need to see me take the place of their old masters and were therefore very concerned about my commitment to the village. I had to promise them time and again that I would not leave until Araouane could stand on its own.

I never ordered the men around or treated them as inferiors, but I sometimes (perhaps unfairly) used their feelings of insecurity to make sure they did their work well. When a problem came up, whether it was slacking off, complaining, a refusal to take initiative or learn new techniques, I would simply say, "I don't want to force you to do things in your village which you don't approve of. I want nothing out of this for myself. If you would rather have me leave, just tell me, and I'll go right away."

Zain, by far the smartest and the best worker of the lot, would usually take the hint. He'd become our new mason, and he remembered a bit of French from his boyhood, when he had served as a household water carrier for his patron in Timbuktu.

"Patron—sorry, I mean, Aebi," he'd say, "please understand us. Many things you, Fritz, or Mohammed Ali tell us, we say we understand, just so that you will not be angry at us. But most of it, truly, we do not. Please do not be mad and leave us. We want to make a good village, we like everything you say. Just sometimes it is difficult because we don't understand."

As the language barriers fell and the gardening routine became more established, our miscommunications became less and less frequent. And still later, when they'd begun to reap the rewards of the project in their diets and their pockets, they began to display a commitment to it for its own sake. I never did tell them, however, that my threat to leave was a bluff all along.

In the beginning only the blacks would deign to work on the project, since they were the ones who did not have an independent source of food. The Arab overlords all had relatives in Timbuktu who periodically sent them supplies by caravan. But there was so much activity going on in town—the work in the garden, the building of new houses, the growing of such luxuries as peppermint for tea, the sense of pride that the workers were beginning to take in their accomplishments—that even the haughtier of the Arabs couldn't ignore it. Araouane was changing, and there was no way they could stop it. Some coped by abandoning the village for Timbuktu, while others pretended not to notice the transformation. They'd sit around idly picking their teeth with little wooden sticks and spitting into the sand.

A young Moor named Dahar, the one who on my first visit had spoken to me in halting French, was willing to work on the project as overseer, but that was just the kind of hierarchical arrangement I did not want. When I told him he'd have to work side by side with the blacks, he went off to supervise black laborers at Taoudenni.

The first real converts among the village "nobility" were Araouata and his little brother Bouama. Araouata was about thirty years old and Bouama eighteen. The two were nephews of Salah Baba and had grown up under their uncle's charge, since their father had died when they were young. When I had first come to Araouane with Dah, I'd spotted an old woman looking out through a small opening in a cubicle standing in the courtyard of Araouata's house; there was no door, just a little opening at shoulder height. Later, I had found out that the woman was the mother of Araouata and Bouama and had gone mad. The sons had walled her in so she couldn't do any mischief, and when she died, they had to break down the wall to take her body out.

When Salah Baba had come with me to Araouane to explain my project, he had impressed on his nephews that I was their last chance. He had virtually ordered them to work in the project on whatever tasks I might assign them. At first they were obviously ashamed to be carrying water, shoveling banco, and making fences alongside the people their family had owned. Neither man had ever done any physical labor in his life, and more than once they came close to collapsing.

Araouata was simply unable to do the same jobs as the blacks. I gave him the undemanding task of checking to see who showed up for work each day, as well as deciding which jobs were suitable for children and which should be handled by adults. Although he still had to carry water as a symbolic gesture, his elevated duties spared him the scorn of the Arab idlers who came to gawk at the workers. But at least it was (I reasoned) a far better object lesson to have him working alongside the blacks, than to make him work so hard he'd have to quit.

Bouama, however, could not be excused on the grounds of physical infirmity. For all of his eighteen years, he'd done nothing but drink tea and gossip. I made him do the same work as all the others, and the black men took a devilish pleasure in giving him the dirtiest jobs. After a few months he ran away to Timbuktu, but old Salah sent him right back up with the next caravan. He had no choice but to do a slave's work.

Three years later, Bouama had become one of the boys. Nobody shoved him around anymore, and once cash wages were introduced, he often worked three shifts at a go. He had grown very proud of his trim, muscular body, and on those rare occasions when he decided not to moonlight, he'd taken to strutting around the dunes in the most elaborate Timbuktu *boubou* that his newly earned money could buy. He'd asked somebody else to bring it to him from the city, because he didn't want to take time off from work to go shopping for it himself. In fact Araouata had perhaps become a bit jealous of his brother; I often observed him struggling bravely at taxing jobs he wouldn't even have attempted in the beginning.

Some of the other Moors were eventually won over by curiosity. After seeing that Araouata and Bouama were none the worse for laboring and had even gained a new respect from me and from their coworkers, some of them decided they wanted to be part of the project. But the greatest influence on the other Moors was certainly Mohammed Ali. He told them of the outside world, how Arabs and blacks worked side by side in other countries, how the situation in Araouane was a relic of a backward time. His hard work backed up his words, and the ranks of our volunteers swelled.

Not all the Arabs started out with a slave-and-owner mentality. When the nomads came to the village, they usually went directly to the house of an old black man named Habbabu, where they could trade salt for ropes, *kitas* (the contraptions which are put over a camel's hump to carry the salt bars and other cargo), or have a meal. It was obvious that they felt more comfortable mingling with the blacks than with their fellow Arabs.

This may have had something to do with the fact that some of the nomads were in a sense servants themselves, to the town's Arab families, who hired them as camel herders. One nomadic aristocrat, the one-legged Sidi, who was reputed to have more camels in the region than anyone else (you'd never actually *ask* a man how many camels he had, just as in America you wouldn't dream of asking someone how much money he has in the bank), spent much of what time he spent in Araouane settling matters with his visiting nomadic Arab bondsmen and his black bondsmen from the salt mines.

Long ago, when the caravans had carried slaves rather than salt to the north, these nomads had certainly been involved in the slave trade. But even then they had generally been transporters rather than owners. Their way of life in the desert wouldn't have allowed them to add to their households—it was hard enough to find pastures large enough to earn a livelihood for their families, and they lived so simply that they had no need of servants.

At the height of the 1972 drought, when most animals in that part of the Sahara had died, a nomad named Abidine had come to Araouane with his last three camels. The rest of his herd had perished, and these animals were very weak, but the villagers were starving. He let the townspeople feast on his camels—his only possessions of any value—and then he left.

"Maybe someday," he said, "you will return the favor."

He left his family in Timbuktu and for seventeen years traveled all over North Africa looking for work. He labored in the Libyan oil fields, paved roads in Algeria, walked the Sahel. Somewhere in his journeys he heard about our project in Araouane. He simply arrived one day, a shriveled old man with one stiff leg who looked like an Italian farmer from the Abruzzi. He didn't ask if we could help him; he asked if he could work.

Of course he was welcome, especially after we heard his story. He showed up on time every day and seemed genuinely grateful for the employment. One day he went off to Timbuktu to find out what had happened to the rest of his family, and we didn't see him again until the next year.

Shortly after that, a young Arab appeared in Araouane. At

first I thought he was a nomad who had just lingered a bit longer than usual and was lending a hand with the garden work. That was not a strange sight, since many of the nomads got caught up in the project for a little while, but it seemed as if every time I looked up, there he was. Finally I asked Mohammed Ali who this young man was, and why he worked without even asking for rations. Mohammed Ali didn't know either and went to ask him.

His name, it turned out, was Sidi Mohammed, and he was the son of Abidine. His father had told him that the family's future lay in Araouane, and had warned him never to ask me for anything because "the foreigner got very mad when somebody begged." So, without a word of explanation to anybody, he had set out to demonstrate his worth by getting right to work. We took him in gladly and gave him his well-deserved rations. Soon he set about building a cute little house. When we started to offer extra after-hours jobs for money, Sidi Mohammed was at it practically morning and night.

One day he came by to ask if, the next time I came up from Timbuktu, I would bring his wife and two children, as well as his father. His house was finished, and he wanted to have his family together in Araouane.

"I have already bought the goats from nomads to pay for their transport," he said. Whenever I drove to or from Timbuktu, I charged one goat for each passenger or 100 kilos of cargo; these goats were then distributed to the workers as special bonuses.

The family settled in Sidi's little house. His little garden was so beautiful that Sidi refused to pick any of the vegetables for fear of breaking up the neat lines of greens, onions, and carrots. His family became valued members of our village, an inspiration to the local blacks and Arabs alike.

The nomad Hamd'r Rahman was obviously considered by everybody in the region to be special: nobler, richer, more influential. When he came into the village, all the young men, black and white alike, would lower their heads to him in def-

erence, and he would touch them in acknowledgment. Once
he hitched a ride with me in the Landcruiser. He had heard
that a son-in-law who lived far to the south was very ill, and
he didn't want to spend a week or two searching for him by
camel.

This was the first time the elderly man had sat in a car, and
he held on nervously for the whole trip. After a few hours of
driving, he saw some nomads in the distance and motioned
me to stop. He knew these men, he said, and wanted to ask
them where his relative had set up camp.

The men came running toward us when they heard him
call. They bowed low to him, banging their heads on the
windows of the Landcruiser—they had apparently never
seen glass windows before and were amazed to find an invisible obstacle. Poor Hamd'r Rahman was very embarrassed for
them and tried to open the window, but it was a feat beyond
him. He fumbled for the correct lever, and before I could
come to his assistance he reached for the door latch. The
door flew open and again banged the nomads in the head. It
was a scene to make Laurel and Hardy proud.

In January of the second year, I helped the people of
Araouane to celebrate their achievements with a spaghetti
dinner at my house. I invited the best male workers and the
Arabs who had given the most help to our project. (At that
time it was still unthinkable that the women could eat at my
house.) The dinner held much that was new for my guests.
Most had never sat at a table, since there was no furniture at
all in their houses. Most had never heard music played from
a tape deck. Certainly none had ever eaten spaghetti. And
rarely had the two groups shared a ceremonial feast.

Since I had issued the invitation, nobody could easily
refuse. But at the outset the groups were ill at ease with each
other, the blacks deferring to the Arabs, and the Arabs trying
a bit too hard to act nonchalant. The atmosphere was stiff, to
say the least. But when eating spaghetti without utensils, it is
difficult to remain formal for long.

The polite way to eat with your hands is to make some kind of ball with the food, holding four fingers in the shape of a cup and rolling the morsel up with your thumb. When the shape is right, you raise your hand to your mouth and flick the food in with your thumb. With long strings of pasta, however, none of this worked. Some men tried to roll the strands into balls, some grabbed handfuls to shove into their mouths, and all ended up dotting their turbans and jellabas with bright red tomato sauce. They looked at each other, laughing, to see who could come up with a practical way of eating this odd foreign food.

It was old Baba Camouse who devised a workable method: Grab a bunch of spaghetti, suspend the strands in the air until they stop swinging, then slowly lower them into the mouth. But by this time all the men were laughing so hard that nobody could keep his mouth open long enough to eat.

The dinner left a feeling of camaraderie among the men. It also left a few dozen red-stained robes, turbans, and beards.

PLANTING THE SEEDS OF THE ROOT OF ALL EVIL

When I first came to Araouane, the villagers had absolutely no concept of money, at least not as we know it. If anybody needed to buy something, he did it with salt. Each piece of salt had a generally recognized value based on its quality and weight. In the days when Araouane was thriving, salt traded pound for pound with the gold that caravans brought in plenty from the south. In those times, of course, Araouanites also exchanged their precious water for goods such as ivory, slaves, rare pelts, spices, carpets, and guns.

At the time of my arrival, the Malian government's official currency (West African francs) had little value in Araouane. If I'd had the foresight to buy a mountain of salt bars for a couple hundred dollars in the Timbuktu market, I could have presented that to the nomads as payment for my passage with the Taoudenni caravan. They were accustomed to buying all their necessities with salt—to them each bar meant so much rice, millet, sugar, or tea. A wad of paper slips meant virtually nothing to them, and the printed numbers on the bills were merely odd symbols that they didn't understand.

Shortly after my arrival in Araouane, I tried to buy a camel from a nomad to feed to the villagers. The animal had a broken jaw which prevented it from eating, so it was useless for anything but butchering. I knew the price would be low. Garba, the mason from Timbuktu, negotiated the transaction. My classical Arabic didn't let me understand the conversation, but I did catch the figure 10,000. When Garba told me the agreed price was 50,000, I thought he was trying to cheat me.

"I distinctly heard him say *ashra*," I said. "*Ashra*, not *chamseen*."

"The man wants fifty thousand," Garba insisted. I couldn't very well call him a liar, so I peeled off five 10,000 notes. The nomad refused to take them.

"He says this is not money," Garba translated. "He wants blue ones, Aebi. Ten blue one-thousands."

Perhaps I wasn't being conned after all. But the 1,000 notes were orange, not blue. I gave him five anyway, and the nomad threw them angrily at my feet.

"Blue ones, Aebi," said Garba.

"The only blue notes are five-thousands," I said.

"Yes," the mason replied. "The nomads do not know any denomination over a thousand. Five thousand, ten thousand, one million—it is just 'one thousand' to them."

I paid the man his ten blue notes—50,000 francs, as Garba had originally said—and he reluctantly gave me the camel.

As for the Araouanites, I soon learned they had an intricate system of IOUs based purely on honor. If somebody needed a bowl, he simply took whatever bowl might be at hand and returned it later. For transactions with outsiders, the medium was almost invariably salt. During my first year in Araouane, every time I went to Timbuktu, the villagers would fill up my Land-Rover with broken salt chunks of all shapes and sizes.

"Can you have my teakettle repaired for this piece?" they would ask.

"Can you get me some thread for that piece?"

"Will you buy me a turban with that?"

I would write down all the requests that I could understand. For the rest Araouata wrote little notes and attached them to the various pieces.

In Timbuktu I unloaded the huge carload of salt at the house of Salah Baba, the only person who could decipher Araouata's writing; when I showed some of Araouata's notes to my Arabic teacher back in New York, he told me that the scribbles bore almost no resemblance to Arabic, Berber, or any other language he knew.

A rich man (in the traditional way of thinking) was one who had vast stocks of salt, or large herds of camels and goats. Even today, a nomad will seldom agree to sell any of his herd unless he is in great need. When I tried to buy goats to feed the villagers, I could barely get the nomads even to consider making a deal; when they reluctantly named a price, it was always nine or ten times higher than the going market rate. Yet these same herders would walk five days to Timbuktu and exchange the animals for a few bags of tea when they happened to run short of that staple. It was the idea of converting goats into slips of paper that could be stuck in a belt that was completely alien to these men of the desert.

When people romanticize the supposedly idyllic conditions of precapitalistic times, I have to smile. Instead of a paradise where greed and want are unknown, I think of a house full of crumbly salt bars. I think of a farmer who keeps his livestock not until it reaches full slaughter weight, but only until he happens to need a new kettle.

Humans have always needed to trade for necessities they could not themselves produce. Without easily convertible currency, this practice is not less important, just more difficult. Grains and vegetables spoil before they can be exchanged, and farmers have to trade animals that are too old or too young to be of much use. A transaction that would be simple in a cash economy grows byzantine.

In spite of all that, I didn't set out with any intention of changing the Araouanites' economic customs. I wanted to teach them how to grow food in the garden, and nothing more. But things don't always work out as expected. The garden itself brought with it concepts of profit and loss, saving, and cold, hard cash.

My first real problem over money led, in a roundabout way, to the death of the village chief.

Mohammed Sultan (that was his name) came out to Araouane on one of my rented trucks to investigate this pro-

ject of mine for himself. He had never lived in the village—
such a life would be far too uncomfortable for a rich man like
him. Instead, he divided his time among his several houses
and several families in Timbuktu. When he arrived at
Araouane, he made a big show of encouraging the people to
work hard so that the project would be successful.

"If you make the garden in my village," he said, after in-
specting the row of fruit trees we had already planted, "I'll
want you to build a house for me right here." It was obvious
that he had no intention of paying for it. In his view, it would
be my tribute to him as a token of my gratitude for his allowing me to help save his town.

Sure, man, I thought, I'll build you a big, big house.

After two days of strutting around the village, collecting
taxes in the form of salt, he demanded that I transport all his
booty back to Timbuktu. I had no objection, since my truck
would otherwise return empty, but I didn't intend to do his
bidding for free.

"How much will you pay?" I asked, hoping I could recoup
part of the enormous cost of the rental.

"We can load only eighty bars," he said, sounding annoyed
at having to pay anything at all, "and the standard rate is two
hundred and fifty francs per bar."

"Okay," I replied, although I knew that a camel caravan
transporting salt from Araouane to Timbuktu generally kept
one in four bars (roughly 2,000 francs, the average price of a
bar in Timbuktu at the time) as payment. But I wasn't in a
mood to haggle.

The truck was going to leave at four in the morning. As I
lay on my bed of millet sacks that night, I thought the deal
over: The truck had brought up 8 metric tons of food, and
yet the chief told me that it could carry only eighty salt bars,
weighing a total of about one and a half tons. I took my
flashlight and poked around the loaded truck. Sure enough, I
counted 320 bars. I saw red! Without waiting until morning, I
stormed into the house where the driver with his helpers
were fast asleep.

"You cheating bastard!" I screamed, beside myself with rage. "I'll see you rot in jail, even if it costs me a million francs to do it!"

The poor driver seemed to have no idea what I was talking about, and the thought dawned on me that this must have been the chief's doing. I decided I could not go after the village headman directly—I didn't yet know enough about the politics of Araouane society. But I could go after him by proxy.

The driver spoke French, and the chief didn't. I explained to the driver that I was going to bring him before the chief and shout my head off at him, but that he shouldn't worry because everything was going to turn out fine. The driver was not happy about the proposition, but I left him no choice.

We went to the chief's house and woke him up. I ranted and raved to him about the swine of a driver who had put 240 more bars on the truck than the chief had authorized. I would go to the police, I yelled, to a lawyer, to the governor himself to get justice, I'd make him pay dearly for cheating and stealing from somebody who had come to Mali purely out of charity. And so on. The driver uneasily translated all of my tirade.

"These black low-lifes can never be trusted," said the chief, using my black driver as translator. Sweating as profusely as the nervous trucker, he promised to take care of everything, to pay for all the additional bars that "the bastard put on the truck against orders." I was really starting to enjoy it all, and had no intention of bringing the scene to a close just yet.

I stomped about and screamed. No way was I going to let the driver get away with it, I bellowed, this was a matter of principle, cheaters had to be punished, I'd see him hang regardless of what it cost me. The poor driver didn't have to make a show of his fear at this point; I resolved then to compensate him as well as I could for his suffering.

By this time the chief had completely lost his haughty composure. He offered to pay even more than the customary transport fee. I accepted immediately, and pocketed the money on the spot.

"But your kindness is no reason for me to keep from going after this cheat," I called as I stomped out of his house. The chief followed me to my own hut. He sat with me on the millet bags and continued our discussion in broken Arabic. He entreated me not to pursue the driver anymore, saying it would not serve any purpose since "these people are all the same." He offered me more money and turned apoplectic when I refused. He would not leave my house until the matter was settled. Finally, since I wanted to go to sleep, I told him I'd think it over.

At 4 A.M. the truck pulled up at my house. Mohammed Sultan got out and asked if I had changed my mind. I turned to the poor driver and raised my fist.

"You'll be in prison," I yelled, "as sure as I stand here! In two days I will be in Timbuktu, and you'll be sorry you ever cheated me!" I couldn't even give him a reassuring wink to reassure him that I was pretending, since Mohammed Sultan was hovering over us like a hawk.

The day I arrived in Timbuktu, the village chief came to find me at Salah Baba's house, where I was unloading salt. He humbly invited me to have lunch at his home. I accepted, and he gave me a meal fit for royalty, but I didn't back off my threat of prosecution. If he hadn't maligned the poor driver so much, I think I would have let him off the hook, but since he'd been so nasty, I decided to make him squirm.

That evening, purely by coincidence, I had dinner at the house of the governor of the Timbuktu region. I didn't bring the chief's case in detail—there were far more important matters to discuss. But word that I had dined with the governor must have gotten back to the chief: A few days later, Mohammed Sultan died of a heart attack.

On subsequent trips, the driver and I became good buddies. Marafat was a great big ambling bear of a man, always ready to smile. I made sure that he was well treated every time he was in Araouane, and he delighted in telling the story

of "our" victory over Mohammed Sultan to everyone there. I
wished he wouldn't brag about it quite so much, since I
didn't want people to think I had respect for traditional authority, but at least it served as a lesson to others. Once a
mine worker tried to con me out of a transport fee from
Araouane to Timbuktu, and when I threw him off the truck,
he asked Marafat to intercede for him.

"Don't even bother," the driver told him. "Everybody who
tries to cheat Aebi is sorry eventually."

Not long after setting up the garden, we started to build a
school and a hotel; it wasn't difficult for me to persuade the
villagers of the need for a school, since there had been
Qur'anic teachers at various times in the past, but the hotel
they had to take on faith. I figured that even if the garden en-
abled Araouane to be self-sufficient in food, the townspeople
could always use hard currency from tourism to tide them
over the inevitable rough times.

I paid the Timbuktu laborers who worked on the construction projects with actual money. The villagers were amazed
to learn that, by sending these pieces of paper home, the
workmen were able to support their families back in the city.
There was still absolutely nothing that money could buy in
Araouane.

The hotel was finished toward the end of 1989, and a
friend of mine from New York came out to be our first guest.
He stayed almost a month, and when he paid his bill, he
counted out the franc notes in front of a crowd of villagers. I
tried to explain how tourism could help the town:

"Julio just paid two hundred and seventy thousand francs,"
I told them. "All this money is for you."

There was no noticeable reaction.

"With this money," I said, "you could buy about one hundred and twenty bars of salt in Timbuktu."

Most of the adults didn't know how much 120 was.

"That is more than all the salt that Sidi brought during the
all of last year."

That got their interest. Sidi had a lot of camels and was always hauling big loads of salt with the caravans.

"So then," one of them asked, "where is the salt?"

The villagers had never quite understood the purpose of the hotel. It had been a lot of work to build, it was very pretty, and people named "tourists" were going to stay there, but the Araouanites didn't really see what was in it for *them*. Most baffling of all, we had built shelves in the rooms and the reception area out of salt bars. True, Aebi had paid for them, but what a waste—who had ever heard of sticking perfectly good salt in banco?

"With this money," I continued, "we can buy forty bags of millet. It takes twenty camels to bring that much millet up here."

They eventually got the general idea, but that wasn't enough. Someday the villagers would have to be completely independent of me and all other benefactors. Eventually they would have to learn how to deal with money on their own. I am no economist, but I decided I'd have to teach them a thing or two about capitalism.

First off, to cut potential problems in the bud, I drummed into all the kids' heads that they should *never* steal. Theft had never been a problem in Araouane before—where would a thief go with his loot? Everybody knew exactly what everybody else possessed. And the very concept of personal property, as we know it, never really existed here. If a person needed a turban, he just took whichever one was available, even if it happened to be on somebody else's head.

Once the first saplings were planted, I started to experiment with the vegetable seeds from Europe. I tried tomatoes, beans, beets, onions, squash, potatoes, peas, lettuce, carrots, peanuts, peppers, peppermint, melons, cucumbers, eggplant, corn, millet, radishes, garlic, cabbage, broccoli, *disma* (a local spice), and parsley.

At first this was all done in a communal plot. Before parceling out the land to the villagers, I wanted to know which crops fared well and which did not, how much water to provide, how much manure, how to protect the young plants from pests. It was difficult enough to make the villagers believe in my scheme, and it would have been doubly difficult if they had to watch their own personal gardens fail from lack of expertise. So we all worked together on the experimental plot, and nobody really understood what I was trying to do.

The first crop ready to harvest was the radishes. They were scarlet, plump, and looked very pretty. I got the children to help me pull them up, then I washed the vegetables, took off the leaves, and called all the workers together to show them what they'd produced. I explained that this was delicious food, and invited them to try some. At first nobody moved, but slowly a few of the kids stepped up and took a handful each. The others realized that at this rate there would soon be none of the strange new food left, so they stampeded up to grab radishes of their own. Soon a full-blown melee developed, with eager townspeople shouting and shoving and joyfully stuffing the strange red vegetables in their mouths.

After the first bite, most of the villagers spat out the bitter radishes. "This is what we are working for?" their looks of disappointment clearly said.

It took a lot more experimentation to discover not only what crops would grow well, but which ones the villagers would actually eat. Onions, sweet potatoes, beets, carrots, hot peppers, and peppermint were received skeptically, until Mohammed Ali and I cooked up special dinners to show the gardeners how the vegetables could be prepared. Squash, zucchini, eggplant, sweet pepper, and lettuce, however, did not go over well at all. And nobody, but nobody, liked cabbage.

Many crops did not flourish under desert conditions. Peanuts, potatoes, and watermelons barely grew to maturity, which is a pity, since they were popular with the villagers. Tomatoes ripened to a beautiful red, but they had very little taste. Garlic would not take root, which a marabout visiting from Timbuktu explained as the will of God, since observant

Muslims must stay away from the mosque for forty days after eating it.

Our growing season was October through early April, the only time when the sun won't make the crops wither. The next fall I allotted individual garden plots to all the families in the project. I let each family decide which crops to grow and was amused by the choices. The former nomads Abidine and Sidi Mohammed didn't much care for any fruit or vegetable, so they chose crops for appearance rather than taste, favoring eggplant and other flora with pretty leaves. They grew neat rows of diligently tended vegetables, but never harvested and ate their beautiful crops.

As people started to take care of their own personal gardens (and spurred on by the cash prizes Fritz offered for the best-tended plots) they began to discover materialism. This was the first time that most of the villagers had anything that was truly theirs. But it took them a while to figure out the difference between having and taking.

Still in the habit of believing that more or less everything belonged to everybody (especially stuff that simply came out of the ground), children and adults alike poached whatever caught their eye. So many people impatiently plucked the tomatoes while they were still green that I almost gave up hope of harvesting any ripe ones. Even old Baba Cambouse got into the act. When caught red- (or rather green-) handed with tomatoes from someone else's garden, he became indignant:

"You were sent to us by God," he told me, "but it is God Himself who makes these things grow. Surely He makes them for everybody."

One of the freed slaves, Humballa, had taken on the job of herdsman. After I started distributing goats to the workers as bonuses for good work, Humballa saw an entrepreneurial opportunity. For the first time, there was enough livestock to merit a full-time tender, and Humballa offered to take care of the village's entire herd if the villagers would take turns tending his garden. But before long he discovered that his neighbors were raiding his plot while he was out in the desert, and he had to give up his job.

A little boy named Hussein, however, took to the new ways with gusto. Some hormonal deficiency had made him unusually small, but his voice was as deep as that of an old man. After tasting macaroni once, he said that he'd be happy to eat no other type of food for the rest of his life. To ensure a steady supply, he planted a handful of uncooked pasta in the ground and watered it every day, an enterprise that earned him the nickname "Macaroni-Tree."

Once I saw Hussein furtively washing carrots in a bucket behind the water reservoir; I asked him why he was hiding, and a big grin spread over his face:

"If I do it when there are other people around," he said, "everybody will want some of my carrots, and there will be none left for *me!*"

It was obviously not all roses. I realized that I had unintentionally introduced the ugly sin of greed. But if I was to show the people of Araouane that hard work, tenacity, and imagination could bring benefits far beyond anything they had dreamed, then greed seemed a necessary evil. If diligent labor did not have a big payoff, why should anybody bother to strive?

I tried hard to let all the villagers start at the same material level, so that they could accrue wealth according to their individual effort. The necessities of life were all provided for: Everybody who worked regular hours on the project (and anyone who couldn't work due to old age or infirmity) received food, clothing, and medicine. To plant the seeds of capitalism I set up a wide range of jobs on the project, so that virtually anybody—regardless of age, intellect, or physical capacities—could be productive. There were suitable tasks for young men, little children, women, and old people.

In addition to the regular garden chores, I made a list of jobs that villagers could perform part-time for actual money: for delivering forty donkey loads of banco to the project from the mud deposits a mile away, a person would earn 300 francs; making a hundred bricks brought the same wage. For hauling in a camel load of *halfa* (the desert straw which we used in construction as well as for animal fodder, ropes, and mats), the payment was 500 francs. Each task that came up was added to the list.

I also told the Araouanites about entrepreneurs. Later on, I said, when a new building would have to be constructed, we would solicit bids from everybody and give the contract to whoever put up the best offer. That entrepreneur would be responsible for providing the food and pay for everyone who helped him, so that the better a man worked and the better he planned, the more he could earn. But all this, I assured them, was quite a way in the future.

After discussing these new opportunities during one of our weekly town meetings, I drove to Timbuktu for provisions. When I came back a few days later, I hardly recognized the project. *Halfa* lay piled all over in huge mountains. As soon as the day's garden work was completed, all the men, women, and children—old and young, blacks and Arabs alike—had grabbed the camels of any nomads who happened to be in the village, promising the bewildered herders God only knows what in return. They'd all taken off together into the desert to bring in load after load of straw, and since there was a full moon many of them worked straight through the night.

Mohammed Ali tried to stem the tide, but nobody wanted to miss out on making as much money as possible; everyone imagined himself being able to afford a whole herd of camels. I called a new village meeting—during work hours, because after work all the townspeople would run off into the desert to gather more *halfa.* I had to impose some restrictions on the free-wheeling forces of the unfettered market. We agreed that *halfa* would only be bought as needed, when specifically requested by Mohammed Ali. To save the town's eleven donkeys from being worked to death, we agreed that bricks would be made only with banco gathered right within the community. Even this solution caused problems: We soon had made bricks of all the banco available in town, so there was none left for use as mortar.

The episode taught me a lesson in basic economics. I found that in order for a free market to function, it must have a sufficient number of variables. With barely forty people working on the Araouane project, we were forced to introduce some guidelines. I was very happy to see, however, how hard all the villagers were willing to work once they were given a bit of incentive.

I could not, of course, pay every person as soon as he'd completed a particular task. Regular food rations (about 10 kilos of grain per person, plus sugar, baobab powder, cooking oil, tinned sardines, or whatever else I could dig up on my monthly trips to Timbuktu) were distributed each Friday, but special bonuses of goats, chickens, and the like could only be disbursed when we happened to have such luxuries on hand. Eventually we set up a system of bookkeeping, and since both Mohammed Ali and I were too busy to manage it, we assigned the responsibility to a boy named Hamma. Although a tall and athletically built kid, he was so shy that he almost never smiled and only rarely spoke. He was the best pupil in our little school, and he managed the accounts superbly. Only two years earlier he hadn't understood a word of French, but now he spoke and wrote it well enough to balance complicated ledgers and keep track of who was owed how much for what job.

Ever since I'd arrived, people had brought me gifts of eggs, chickens, and camel milk; this last item was the traditional payment of the nomads, who let the villagers milk their camels in exchange for the use of the town's wells. Once the garden was firmly established, I told the townspeople that I no longer wanted them to bring me presents. I told them that when I needed any of these luxuries in the future, I would pay them for it.

I set up a scale of prices for all the goods I might need, and made them considerably higher than those in Timbuktu. Someday, I reasoned, when the villagers would have to purchase goods in the city, they'd have to pay the additional costs of a local agent and caravan transportation; I wanted them to get used to higher prices when I was paying so that they wouldn't be shocked when they had to pay themselves.

Well, the people learned fast. I only had to say the word "eggs" and every village woman raced to close the deal. They even grabbed the offspring from fertile hens, and when I cracked the eggs open, I often found fully formed chicken embryos inside.

The first time I put out the word that I'd like a plucked chicken, a virtual bloodbath ensued. In no time at all my house looked like the poultry counter at a large supermarket. A cloud of feathers drifted in the wind all the way from the town center to the outlying huts. From that point on, whenever I wanted a chicken, eggs, or any other luxury, I had to choose one person to ask.

"Don't forget," I'd often be scolded, "next time you need a chicken, it's *my* turn."

Every year before returning to Araouane from New York, I assembled my provisions at my brother Peter's house in Appenzell, Switzerland. Virtually all the supplies and tools for the project came from Switzerland, and a cousin of mine who is a doctor rounded up boxes of medical supplies. Each year my relatives collected used clothing from their neighbors to distribute in Araouane.

One of my brother's neighbors makes delicious goat cheese. I decided to introduce a cheese industry to Araouane. But since Saharan goats are generally too scrawny to produce enough milk for cheese, I figured we'd have to use camel's milk instead. I got my brother's neighbor to explain the basics of cheesemaking and to show me where to get the necessary implements, although she was doubtful that the plan would succeed. My brother, who is a food researcher for a major food and spice producer, searched the research data and came up with reports of half a dozen failed camel cheese experiments from China, Egypt, and elsewhere. Still, I was determined. When I returned to Araouane in 1990, I brought all the equipment necessary for dairy fermentation.

The first batch of camel's milk cheese we made in Araouane was successful, but only barely. We only got about a fifth as much cheese per gallon as milk from a healthy goat would have produced. But the villagers liked the watery residue far better than pure camel milk, so neither the cheese nor the by-product went to waste. We tried all kinds of different temperatures, different amounts of yeast, different cooling periods, milk directly from the camel's udder and milk left to curdle. The amount of cheese per volume of milk never increased much, but we were able to improve the taste with salt, herbs, and spices.

I began to have visions of exporting our cheese throughout the Sahara. The villagers could drink the whey and sell the cheese itself at the market in Timbuktu.

But when we tried to store the cheese for transportation, we got a nasty surprise. It quickly turned from the consistency of Brie to that of wood. After two weeks it could barely be cut with a rasp, and a few days later the chunks of cheese could have served as hammers. But we kept on experimenting, and for that we needed a steady supply of camel's milk.

In the forty-two years prior to my arrival, there had been not a drop of rain in that part of the desert. As a result, the entire region around Araouane had no pasture at all, and no one but the nomads owned any livestock. And because the cargo caravans only used male camels, for a long time there was virtually no camel milk available at all to the convoys. But the rain returned in 1990, and it brought back the pasture, which in turn brought back the herds. Now there was more pasture available near Araouane than down by the Niger River, and nomads drove their whole flocks toward us. All of a sudden, there was so much camel milk available that at certain times the townspeople could not have drunk it all even if they'd tried.

Once I offered to buy the milk, we were flooded with buckets and buckets of it. The nomads must have been amazed at how eager the townspeople were to milk their camels day and night, and to pay well for the privilege in goods and services. We built a proper cheesery, and Araouata was appointed official cheesemaker. (He quickly became so skilled that Fritz wanted to teach him how to yodel!) We marketed the cheese—as "camelbert"—to the European aid workers in Timbuktu.

All of this had some fortunate side effects: The nomads now were so warmly welcomed that they started to come through the village far more often than before, which provided us with a rich abundance of camel dung. With fuel now easily available, the women of the village did not have to spend most of the day roaming the desert in search of a few turds to burn for the evening meal. They had more time to be productive at home and in their private gardens. They washed clothes more often, and the town started to look much neater and cleaner.

Once the hotel opened, people had an even wider array of opportunities to make money. The kids who served the hotel guests got paid; all the families who wished could take turns cooking for tourists for pay; Fritz had promised cash awards to the keepers of the best gardens, and I offered eventual prizes to the people who built the nicest houses on the garden's periphery. This was not mere magnanimity—these houses could eventually be used to lodge guests once the hotel was full and would, in any event, serve as a wind shelter for the delicate crops.

With all this money floating around the village, I had to find a way of teaching the townspeople how to manage their finances wisely. We had many discussions about guidelines, but on one point I was firm: In Araouane nobody would give or receive credit from anybody. I forbade Mohammed Ali to permit anyone to run up a tab at the general store. I'd seen too many of my friends back in America—people who ought to have known better—get into too much trouble with credit cards.

Payment for after-hours jobs was relatively small; income from extra work was gravy. I tried to impress upon the townspeople that I would be very disappointed if they used their extra money to buy useless luxuries. I explained the difference between capital investments and mere consumer goods, using the example of a camel and a radio. A radio, I pointed out, would give some instant gratification, but would soon enslave its owner to the constant demand for batteries and would produce nothing of value. A camel, on the other hand, could also provide instant gratification (you could slaughter it and feed your whole family for weeks), but if tended instead, could provide milk indefinitely, and even produce a succession of baby camels in time. A radio, in other words, would only drain their resources whereas a camel would multiply their wealth.

When the villagers saw the Timbuktu workers occasionally giving me money to buy bouillon cubes, tobacco, or rubber sandals on my periodic trips to the city, they wanted such luxuries as well. With their newfound wealth, the list of requested items seemed endless. I didn't want to be bothered doing the whole town's shopping every time I went to Timbuktu, so I started bringing supplies of the most popular articles back with me to keep on hand in Araouane. Soon this grew to a full-blown general store, and I turned this over to Mohammed Ali.

Mohammed Ali's store sold little beyond the bare necessities of life, and that was just fine. At the beginning of our foray into a cash economy, people would sell me eggs and immediately turn around to exchange the money for bouillon cubes. I lectured the townspeople a great deal about investing, but I also believed that they should have some way, however small, of enjoying the benefits of their hard work.

Gradually, therefore, we expanded the stock to include a wider range of items. Unscrupulous traveling traders and even more unscrupulous "patrons" had traditionally provided minor luxuries at ridiculously inflated prices, so Mohammed Ali and I decided to offer them at a fair price to let the villagers save their money. We began stocking the store with cloth, tea, sugar, razor blades, and Vaseline (which the townspeople used to salve their horribly cracked feet; whenever they couldn't get it, they used motor oil instead). Soap had become a big item, which helped make Araouane one of the cleanest communities in North Africa. Clothes got washed, children got scrubbed, and soon I didn't have to hold my breath when I stepped into a packed room.

Mohammed Ali's store became a roaring success with the nomads of the surrounding wastelands as well. These shy desert dwellers were now able to procure all their necessities at a friendly little trading post right in their own territory instead of having to make a long voyage to the big city of Timbuktu. They had no money at all, of course, so they had to conduct all their transactions by barter. They traded camel milk for macaroni, couscous, flour, rice, peanuts, and dates— all items that the villagers got as standard rations for their garden work. Mohammed Ali traded all these items for pieces of salt, and often gave change in salt as well. Soon his collection of salt chunks would not fit into a money box, and we had to build a warehouse for him to hold his treasury until it could be sent to Timbuktu to buy more goods.

Before long, we had become the main topic of conversation at campfires all throughout the region. One man named Habbabu used to take visiting nomads on tours of the project, showing them all the marvels we'd created. This led to some hilarious encounters: The nomads are unaccustomed to walls, of course, so sometimes they would get confused in the houses and enclosures of Araouane. Three or four of them would flock together like frightened sheep, running about trying to find a way back to the open space while the local kids watched and laughed. One time I came into my house to find three nomads huddled in a corner—they'd been there for hours, I was told; when the screen door had swung shut behind them, they'd thought they were in a trap and couldn't think how to get out.

Since Mohammed Ali had had no money to provide the store's original stock, I'd bought the first shipment for him in Timbuktu and Bamako. Later on smugglers, who bought state-subsidized staple foods in Algeria and sold them for a profit in Mali, discovered our market and started selling their goods to us at a discount to save the trouble of seven more days' traveling to Timbuktu. Until he made back his advance, Mohammed Ali paid half of his profits to the Araouane fund, a sort of savings account that I'd set up to provide for the village's financial needs after I left. Eventually the fund would be large enough to buy those few necessities the town couldn't produce or barter, and Mohammed Ali would own his store free and clear.

In one of our last weekly town meetings I explained to the people that I hoped Mohammed Ali would become a rich man. I hoped that he would be a wonderful example of how a man could come to the town possessing nothing at all and (with a bit of help in the beginning and a lot of hard work along the way) could become wealthy without lying, cheating, or stealing. I told the villagers to observe their teacher: If he was successful, they all could be too.

Toward the beginning of my third year in the village, the men whose freedom Fritz and I had bought asked for a special meeting with us. They looked rather somber, so I was apprehensive. What could have happened now?

Mohammed Hassan spoke first.

"We have some debts," he said, "and the people who have given us the credit want us to work for them."

"Who has new debts?" I asked.

"Faradji and I."

"How did you manage to get into debt?" I demanded. "You get all the food, clothes, and medications you need. Your

A salt caravan arriving in Araouane from Taoudenni.

A caravan at the wells; Araouane is in the background.

3

I get comfortable with the nomad women in Dah's camp.

Women of Araouane, dressed up for Dah Ould Sultan's wedding.

Hacking out salt in Taoudenni.

The women and children prepare banco to make bricks.

Digging up banco from
under the sand.

Little helpers.

The beginning of the project; the trees are enclosed with
straw mats from Timbuktu.

The garden two years later, photographed from the same spot.

The courtyard of the hotel, with American guests.

The school and the hotel, seen from the garden,
with Fritz's splendid facade.

School in session.

Mohammed Ali teaching the children the names of the vegetables
and how to prepare them.

Araouata and Bou-djema with produce from the garden.

Amma in her wedding dress.

Babaya and his wife Kia and some of their children.

Wise old Baba Cambouse, who remembers French from the time
of the Foreign Legion.

My truck stuck in the sand, with guides Moulay Mokhtar and Ali.

Emilie scraping a gazelle skin, just before we had to leave
Araouane in March 1991.

Ernst with a load of food from a smuggler's caravan.

children get free education. You live in your own houses."

"I got a pretty dress for my wife," said Mohammed Hassan.

Both men smoked pipes all day long, and I'd often wondered how they were getting the tobacco, which was not included in the rations.

"To whom do you owe money?" I asked.

"Our old patrons came through on their way to Taoudenni."

We all looked at each other in silence for several minutes.

"What do you propose to do now?" I asked.

"We have to go to Taoudenni to cut salt, to pay back our debts."

"Do *all* of you want to go back to Taoudenni?"

Zain spoke up. "You know, Aebi, we are very grateful to you. Everything here is better now, but sometimes we would like things that we cannot get if we don't have salt. In Taoudenni some of us, if we work very well for the patron, can work two days a week for ourselves. With that salt we can buy tobacco, sugar, tea, even some nice blankets from the Tuaregs. And whenever we need more money, the patron gives it on credit.

"We know that you don't like to give credit, and that is all right, but sometimes . . . you know, it would be nice sometimes to get something special."

Only two years ago these people had struggled simply to survive, and now this! But the more I thought about it, the happier I was about the development. These former slaves had gotten a sense that with more work (and no credit—that seemed to be the key) they could get more of whatever would make them happy. Here was the initiative and drive that could keep the project running without any outside push! Once set on the right track they wouldn't need anybody telling them when, what, or how much to do. After a long and amicable discussion, we came to the following solution.

Mohammed Hassan and Faradji, because they had already gotten into debt and therefore assumed the responsibility of going back to the mines, would leave with the next caravan. One half of what we had paid for their freedom they would still owe to Araouane; they would try to repay this sum as soon as possible, because the longer they stayed away, the more tempted they might be to buy more goods on credit. They wouldn't have to cut much salt to pay off their existing debts.

We waived the other half of their freedom price—they could consider it payment for having worked for a year and a half on the project. They forfeited their claims to a garden and house on the project land (there had to be some penalty for defaulting on an obligation), but their children and wives would still get free education and medicine. I saw no need to punish the families for the foolishness of the breadwinners.

About four months after leaving for Taoudenni, Faradji sent 25,000 francs back with a caravan. It was almost half of his debt to Araouane. He must have worked extremely hard, and how he converted the salt to cash is still a mystery to me. But it showed just how much he had learned on the project: The same man who had let himself fall deeper and deeper into debt every month was working to pull himself out of the hole.

From the beginning I'd drilled into the children's heads that they shouldn't get spoiled when visitors would give them presents. They knew that it was bad to beg—begging had become a dirty word long before there was even anybody to beg from! The children all proudly recited a motto I'd taught them: "We don't lie. We don't cheat. We don't steal. We don't beg. We do the best work we are capable of doing." When I still was the town's teacher, those were some of the first phrases the kids learned. They were very proud of their motto and kept watch on each other to make sure it was enforced.

Later on, when tourists started driving to the village in their all-terrain vehicles, the villagers amazed them with their honesty. One Swiss couple who stayed in our hotel gave me an enormous bag of chocolates, pen knives, pencils, and T-shirts. They told me they'd expected to dole out these goodies when kids pestered them for *cadeaux* on their trip. But all the time they'd been in town, no one had so much as hinted at a handout, so they wanted the kids to have the lot.

Most of the tourists who stopped in our village had driven through so many thief-filled settlements that they had grown paranoid about townspeople's intentions. The children would indignantly inform them that they did not need to lock the doors to their cars or hotel rooms, that they could even leave their wallets in the middle of the village and find them untouched the next day.

"This is Araouane," the kids would explain proudly.

POLI SCI IN THE DESERT

Just as I hadn't planned on transforming Araouane's economy, neither had I intended to alter its political structure. All I wanted was to show the locals how to plant trees and cultivate a garden, not to reweave the fabric of their society. Indeed, for a while I had only a vague idea of how local politics worked. There seemed to be a rather strong social hierarchy, but though the *whos* and *whats* were clear, the *whys* and *hows* eluded me completely.

I was able to get a few hints from the village's architecture. Some people lived in big, relatively sturdy compounds surrounded by high walls, while others had only isolated mud hovels. The way the villagers dressed also told a story. Some of the Moors wore well-tailored (if well-worn) jellabas, but much of the population dressed only in rags. Many of the women wore sheets of cheap Chinese fabric, which in other parts of Africa are commonly used as tablecloths. The smaller black children generally wore nothing at all.

Though I tried to stay out of social policy, it soon became clear that there could be no absolute dividing line between the running of the garden and the running of the village. The garden overshadowed every other activity in Araouane, so I couldn't really pretend that I "only" controlled my own little project. Every decision I made had a direct impact on all the villagers. I knew I had to find some way of bringing them into the administrative process. I knew that whatever system I set up would not only have to be culturally acceptable to the villagers, it must also eventually enable the people to govern themselves effectively without any outside direction.

I didn't think that my degree in political science from Zurich University was going to be of much use, but I resolved to try.

Traditionally Araouane had been a society separate from any of its neighbors, and these days it was completely ignored by the national government. Its people did not even speak the language of the nearest city. In short, Araouane was a perfect laboratory for political experiment. With more than a bit of arrogance, perhaps, I started having visions of realizing mankind's age-old dream of utopia.

Culturally, the people of Araouane seemed best suited for a communal system: There were no strong traditions of individual rights and responsibilities, and they were accustomed to think of themselves first and foremost as members of a social group. So I tried running the village like a commune, with everybody working together toward the common goal of a rich harvest.

None of the garden tasks fit easily into the villagers' pattern of communal activity. All the project work was so unfamiliar that they seemed to regard it as a lark. When I distributed watering pails, for example, the villagers kept them for their personal use. It wasn't that they lacked civic-mindedness, they just didn't see why they should use perfectly good pails to pour valuable water onto bare ground. Even more idiotic, in their view, was my insistence on making soil beds from camel dung. This valuable commodity was the villagers' only source of fuel, so my request that they toss it in the ground, water it, and leave it to rot must have seemed like madness. Not surprisingly, my requests for voluntary contributions usually fell on deaf ears.

Once I drove a bunch of women and children in the truck out to the desert, to a site where a huge caravan had recently camped, to gather more dung in a day than they could have found near Araouane in several weeks. We filled the truck and made a triumphant return to the village, the women and children sitting proudly on the huge pile of dried manure. But when I asked them to throw it all into the seed beds, I faced a full-scale mutiny. My assurances that it would all be for the good of the town were in vain. Only when I let each of the collectors fill a millet sack with dung for his or her own use did they agree to help me fertilize the garden. I could only imagine what their dinner conversations were like that night.

It became clear that we were a commune in name only. The people were not truly running their own lives, but merely letting me do the running for them. Far from being a perfect democracy, it was a benevolent (I hoped) dictatorship.

To be truthful, I had always thought this the most efficient form of government, provided the dictator truly sought to benefit the people rather than merely to enrich himself with their labor. My experiences on the board of my SoHo co-op in New York City had confirmed this opinion: A dozen intelligent, well-educated people regularly sat down to try to run the apartment building, and we couldn't even decide what color to paint the stairway handrails. Years later, when I owned my own building and called all the shots, everything ran smoothly and without complaint from the tenants.

Since I was confident of my own altruism, I had no qualms about being Araouane's autocrat for a time. The ultimate goal, however, was to make Araouane independent of outside help, so eventually the townspeople would have to learn how to fend for and govern themselves. Most villagers had never had to make decisions about their lives before: The workers had merely followed the orders of the patrons, and the patrons generally lived in Timbuktu. Although this had instilled habits of obedience that helped to set the project in motion, it would not sustain the town once I'd left. So I tried to delegate as much responsibility to the villagers as I could, little by little helping them to take over the running of their lives.

To inculcate individual spirit, I divided the communal garden into separate plots and parceled them out to all the villagers who were actively involved in the project. I made certain that the townspeople knew that owning a plot, like being a citizen, was a privilege that carried with it civic responsibilities: I refused to allocate any land to people who had not already demonstrated their social commitment by contributing to the communal effort.

The responsibilities of the garden owners were clearly spelled out: If the plot was not productive after a year, it would be taken from its owner and divided up among those farmers who had tended their parcels with greater diligence. Those villagers who made their gardens flourish would own the plots free and clear after one year. After that, they could even sell their land, but I hoped that most of them would continue their agricultural work. Every family that had at least one member working on the project was given a parcel, as was every single woman. I continued to supply equipment, seeds, and rations after the breakup of the communal garden into smaller pieces.

At first Fritz and I offered cash rewards for the best-maintained gardens to stir up the spirit of competition. Once the villagers saw that harder work would bring them greater income and a more comfortable life, each farmer labored to make his or her plot the very finest. They not only built protective walls to shelter the young saplings from sandstorms, but on their own initiative built purely decorative walls as well. It was not unusual to see whole families busily tidying up their private plots long after the official workday was over.

Before long, evening discussions began to revolve around agriculture rather than idle gossip. Townspeople debated whether to plant extra crops of sweet potatoes or tomatoes or carrots, exchanged advice on irrigation and seeding techniques they'd discovered. They became an example of Adam Smith's theories in action: Capitalistic division of labor arose naturally, with each member of society concentrating on the tasks at which he excelled and exchanging his labor or expertise with members who possessed complementary talents. A big child would build a banco wall around the plot of a kid too small to carry the bricks, and in exchange the younger child would water his friend's plot for a week. Men would provide heavy labor to women in exchange for water carrying or weeding.

One day I surprised Boudj drawing plans for something on my kitchen table.

"What's that?" I asked curiously.

"A teahouse for nomads."

"And where are you planning to build it?"

"In the old chicken coop."

"Do you suppose the nomads are going to pay for tea?" I asked. The laws of desert hospitality dictate that any guest is treated to as much tea as he can drink.

"No, but my neighbors will pay me to keep the nomads here for as long as possible. While they are drinking tea, we can borrow their camels to get *halfa*. And maybe the nomads will let me milk the camels as well."

We had many discussions about how the villagers could carve out individual niches for themselves, and how the village could eventually become self-sufficient, but most of the conversation was between me and the children who hung around my house most evenings. I wanted the whole village to take part in this process, so I decided to build a public square where regular gatherings could be held.

Mohammed Ali had the idea of using drama to help communicate whatever situation we were discussing. Fritz designed what he imagined a Moorish amphitheater to look like (if such a thing existed). We built it on the garden's periphery and dubbed it Le Théatre. From that point on, every Thursday, one hour before the official work week was over (we took Fridays and Saturdays off in deference to Islamic tradition), after we had cleaned and stored all the gardening tools, we gathered for a town meeting.

Mohammed Ali and I would go over what had been done well during the week and what could be done better. Anyone else could speak out on any subject. At first it was difficult to get the villagers to speak up, but gradually, with old Baba

Cambouse leading and Babaya Ould Sultan (the brother and proxy of the current village headman, who couldn't be bothered to leave Timbuktu) not far behind, the others would get into heated discussions. As a respected elder, Baba Cambouse was a great help in conveying the urgency of our plans. Without his eloquent support, I might never have gotten the villagers to claim the garden as their own. And it didn't hurt when he got Babaya to give his stamp of approval.

Of course, this was still far from laissez-faire economy. We still had a great deal of central planning and community support. But within a few months most of the villagers were participating in all aspects of the planning.

I had initially been worried that the existing social hierarchy would make egalitarianism impossible. Would lordly Moors consent to take orders from black laborers? Would they even consent to work side by side? Fortunately, my fears were misplaced. The work was so completely different from any work the villagers had known before that Mohammed Ali and I were able to assign tasks on merit rather than social station. An Arab might have felt it beneath his dignity to cut salt alongside his traditional servants, but there was no age-old barrier against raising vegetables. The array of new jobs available also allowed blacks to fill leadership roles for the first time in many years.

Normally, Babaya would have directed any major new village undertaking. He was a tall, bearded fellow who looked a bit like Sean Connery. He had a wife, two concubines, and five children in Araouane, who all lived together in the same house, and in Timbuktu, I heard, he had two other families whom he visited from time to time. He was very smart, and I couldn't help liking him even though, every time we shook hands, I was tempted to count my fingers.

But Babaya knew absolutely nothing about building techniques, and a former slave named Zain had become a proficient mason over the past few months. Babaya acceded graciously to Zain's authority, knowing he hadn't the requisite expertise, and Zain's inherent dignity prevented potential

frictions. He never flaunted his newfound authority and re-
stricted it exclusively to matters related to his work.

Not that the social transformations were entirely painless.
Some of the Arabs, since they'd never had to do any hard labor in their lives, were too weak to undertake any of the
more physically strenuous tasks. They showed their spirit by
working full hours, but often they were qualified only for
jobs traditionally alloted to women. One such task was carrying water from the wells to the garden or to the construction
sites. From the time they are little girls, all Araouane women
learn how to balance huge buckets of water on their heads.
This method takes the strain off their arms, but it is not easy
to master. Often the Arabs ended up drenched, to the great
amusement of the women and the blacks.

As the Arabs began to work alongside their former slaves
in the garden and on village projects, they had a more difficult time getting the blacks to perform their traditional services. It had long been expected a black would do any
household chore demanded in the home of any member of
his or her patron's extended family without payment; it was
enough that the patron supplied sustenance. But once they
were receiving full food rations for their labors, the blacks no
longer needed to rely on the Arabs for sustenance.

In the spring of the third year, a British photographer, John
Evans, and his associate, the writer Fleur Levene, passed
through Araouane on their way to the salt mines. They
wanted to arrange a camel caravan, and instead of finding a
nomad we decided to charter the trip from the village.

Three of the Arab aristocrats owned enough camels, but
only Babaya's herd was close enough to serve. He negotiated
the price of six camels and a guide to Taoudenni and back,
and calculated that he would make the princely sum of about
$2,000 from the enterprise. All he had to do was get his
camels from their pasture grounds two days away.

He told Habba, another of the slaves whose freedom Fritz
and I had bought, to go get the animals, a request Habba
would have jumped to fulfill a year or two before.

But now he said, "I am sorry, Babaya. I cannot go. I am working with the brick-making team now, and if I left, I'd lose several days' pay."

Babaya obviously couldn't believe his ears, but he was so excited about the prospective deal that he didn't get mad. He went to Habbabou, a black man who had been the chief provisioner for nomad cameleers. Habbabou refused Babaya's request as well. Some nomads had asked him to make a lot of rope for them, he said, and they'd already paid him in salt.

It never occurred to Babaya to pay his former bondsmen for services that traditionally had been rendered for free. He had no choice but to trudge out into the desert himself. He came back with the camels, and needed a lot of help saddling the animals up and preparing them for the journey. Habbabou gave him a hand, in between stretches of rope making. When the foreigners paid, Habbabou sidled up expectantly, obviously assuming he would get a share. But Babaya kept the roll of bills tight in his fist.

I urged Babaya to pay the villagers, including his former slaves, for their work. If he was stingy with them, I pointed out, the next time a deal came up, they would see that it went to more generous patrons. Araouata agreed, promising that the next time anyone needed camels he would reward anyone who helped him with the provisioning.

Babaya got the point. He peeled off some blue bills and handed them to Habbabou, who took the notes with a grin.

The former lords of Araouane were well aware of the changes taking place in their village, and many of them deeply resented it. Araouata, still under orders from his uncle Salah Baba to cooperate, reluctantly went along with all our reforms. But other scions of the merchant families, especially Babaya and Hanta, tried to fight change with every weapon they could think of. They didn't dare make a frontal attack, since I had not only the townspeople and Salah Baba but also the provincial governor on my side.

Lamine Diabirra was a Bambara from the south who had

befriended me in Timbuktu. He had been educated at a military academy in the United States and was a passionate admirer of all things American. On the scale of eternal achievement, he ranked John F. Kennedy right after God. I had an open invitation to his home whenever I was in Timbuktu, and he often got annoyed when I had dinner at other peoples' residences instead.

Diabirra shared my contempt for bureaucracy and red tape, and quite openly admitted that government salaries were so low that officials were practically forced into graft, extortion, and embezzlement. He must have taken this admission a bit too seriously, because during my fourth year in Araouane he was jailed for trying to overthrow the government and for excessive corruption. But while we were setting up the garden project, Lamine Diabirra was a powerful friend.

The local Arabs were in cahoots with the clique of Timbuktu merchants who had been milking Araouane for years. The village headman and other nobles were angry at losing the income they'd been siphoning from charitable aid for the town. They constantly plotted to sabotage the project.

Little by little I learned how the clique of Moorish Araouane merchants living in Timbuktu had been milking their ancestral village. For years they'd appeal to all sorts of international relief agencies, describing the sad plight of their town and citing a starving population of 3,500 (nearly twenty times the actual figure). Often they would receive food, clothing, and even cash, none of which ever reached the villagers. Toward the end of my second year I happened to see an application these merchants had made to the United Islamic Fund; in it they asked for the equivalent of $150,000 as reimbursement for personal efforts they had supposedly made on behalf of the village. The application went on to describe, in perfect detail, the project *I* had started—and financed!

As the project became better known throughout the region, their game became increasingly difficult. Many of these

cheats had lived exclusively on the income from such schem-
ing. I blew the whistle on them wherever and whenever I
could. Babaya and his cohorts had no official positions, but
since they'd been the traditional bosses of the village for
many generations, the Malian bureaucrats considered them
quasi-official authorities. They made good use of this fact,
and as a result, despite my friendship with the governor, I
sometimes came into conflict with the national government.

In February of 1990, my second year, an old Land-Rover
with six heavily armed gendarmes (the national police force
of Mali, as distinct from the local squads) arrived in Araouane. The children came frantically running down to the
project to tell me, and I returned with them to greet our visitors. One of the officers was the regional gendarme chief, a
black man in charge of all of northern Mali. The other officer
was a Tuareg from the Timbuktu garrison. They barely ac-
knowledged my greeting, saying they would meet with me
after they'd spoken to the townspeople. They left two of the
guards with the vehicle, while the other two accompanied
them around the village.

All evening the children brought me reports of the officers'
activities. They were interviewing all the older villagers. Fortunately the black gendarme commander did not limit his
talks to the Moorish aristocrats, as any Arab would have. The
police took one townsperson after another into Babaya's
house, which they had turned into an interrogation room.
The children reported that they had kept old Baba Cambouse there for more than two hours. Some of the town's elders came to see me after their questioning was over, and
said they'd been grilled about every aspect of the project:
who did what jobs, how much they were paid, who was in
charge, what was their opinion of Aebi, the list went on and
on. The old men were deeply worried. The interpreter,
moreover, was a stranger from Timbuktu whom the villagers
did not trust. I told them that as long as they'd told the truth
there, was nothing to be concerned about. We were doing
nothing wrong, so no harm would come to us. I tried to reas-

sure them, but I wasn't sure I believed myself.

In the morning, the gendarmes called some of the men they'd already spoken to back for further interrogation. I was glad that I'd counseled everybody to tell the whole truth; if anyone started holding back information in hopes of protecting me or the project, the police would certainly discover it and conclude that we had something to hide.

In the afternoon I invited the gendarmes over for tea. They declined and demanded to see the garden instead.

I decided to let my irritation show.

"Around here," I said, "we treat each other with respect and courtesy. I would appreciate it if you would do the same. For starters, please tell me your names."

That seemed to loosen them up just a bit. They introduced themselves, and I showed them all around the project. Wherever there was a locked door, like those on the storeroom for the school supplies and on Mohammed Ali's little store, the officer demanded that it be opened immediately. Mohammed Ali was off in Timbuktu, learning first aid at the hospital, so the boy Boudj had been left in charge of the keys. His hands were trembling so badly that I had to open each lock for him. We must have looked guilty as hell to the gendarmes, but they searched every closet and shed without finding anything suspicious.

After examining the whole project, they asked me where we could have a talk. At this point I did not feel inclined to invite them to my home, so I took them to the dining room of the hotel. The four policemen sat down (the other two were still guarding the Land-Rover) and spread a sheaf of papers on the table.

"Mr. Aebi," said the commander, "we have many reports about you from the capital. It seems that you flout all the laws of the Republic of Mali. This report from the Ministry of Transportation states that you conduct a taxi business without any kind of license. It says that you have a 4x4 truck and transport goods and people between Timbuktu and Araouane, for pay."

"This letter from the Ministry of Tourism," his assistant added, "states that you run a hotel in Araouane without any permit. Here we have a report from the Ministry of Communications that you operate a radio transceiver without the proper clearances. The Ministry of Internal Affairs reports that you work in Mali without a work visa, and that you completely disregard the rules of appropriate village authorities. From the Customs Department we have reports that you've brought all kinds of agricultural material and equipment into the country without paying the required duties."

They shoved the stack of papers over to me. "How do you respond to all these charges?" the commander asked. "Are they true?"

I glanced at the papers. They seemed genuine, all on official stationery and covered with lots of stamps, Ministry of this and Ministry of that, and notations that duplicates had been forwarded to some office or other, not to mention the curlicued signatures typical of Malian bureaucrats.

I shoved the papers back to them.

"I guess that most of this is quite true."

"So how do you propose to deal with it?"

"I don't propose to deal with it at all," I replied. "If I were to comply with all that stuff, I might as well forget about the project entirely. I am not equipped to bother with bureaucratic requirements. I have no office, I have no secretary, and from Araouane there is no way of communicating with the outside world except by private transceiver. I have neither the time nor the inclination to sort through the mountain of paperwork you suggest, and the villagers can't do it either. At least, not until we have taught them how to read and write—something *your* government has never bothered to attempt."

They were taken aback by my bluntness, but I continued.

"I do realize that Mali is an independent country," I said, "and I fully respect your laws and regulations, but under the present circumstances, I am in no position to comply with them. You have visited the project, you have seen what we do, you have interviewed all the villagers and found that they support the plan. If not for my efforts, this town would prob-ably not exist anymore and its population would add to the hordes of refugees camped out in Timbuktu.

"You ask me how I propose to deal with the accusations? Well, all I propose is to keep doing what I am doing, and if you don't like it, just say the word, and I'll pack up and leave."

My speech brought only a stunned silence. They'd clearly expected some sort of hurried excuses or a plea for compro-mise, not an ultimatum. After all, these were the all-powerful gendarmes, members of a force that answered only to the president himself.

"What's more," I added, "the people you *ought* to be inves-tigating are the very ones who reported me to you. I know that the government could not have found all this out on its own—there hasn't been a state representative here in years, and the town has been entirely abandoned by the central au-thorities. The late headman, Mohammed Sultan, used to col-lect taxes in salt from the starving people, but I very much doubt that any of the revenue ever reached Bamako. It is that pack of thieves in Timbuktu who have tipped you off to me, that same crooked bunch who have been using Araouane to cheat both the government and international aid donors for decades.

"Just so you don't have to use those lowlifes as spies in the future, let me tell you a few more transgressions of which I am guilty: My car is just as unregistered and illegal as my truck. I operate a school without any certification or ap-proval. I have undertaken construction and well-reclamation without even asking for a permit. The list goes on and on. I have nothing to hide, and I keep the provincial governor in-formed of all my activities, but I simply cannot allow myself to get bogged down in red tape."

"What will we do about our report?" the commander asked.

"I have no idea," I said. "But I suggest you merely explain what you have seen here, inform the ministries of my position, and then tell me what decision they reach. I will stay or I will leave, but I won't change my ways."

As I spoke, the atmosphere was tangibly thawing, and before long we were sipping tea together. We chatted about my hopes for the village, about local politics, about my running battle with the Arab merchants in Timbuktu. There was obviously not much love lost between the black gendarmes and them either. By the time they left, the commander assured me that they would do everything in their power to keep the central government off my back.

"Under no circumstances do we want you to leave," he told me earnestly.

His men wanted some postcards of Araouane which I'd had printed for the tourists we expected. That's probably illegal too, I said to myself with a smile. I made the gendarmes pay full price.

Later, the villagers questioned Boudj, who had been in the room during the proceedings, about what had happened.

"They asked him for papers," he said, "and Aebi told them that they had enough already."

After that I was hassled by the government only once, when the Ministry of Tourism tried to force me to register the hotel. Perhaps they'd seen the postcards and gotten jealous. I gave the summons to the governor to deal with, and was never again bothered by any bureaucrats until my last days in Mali.

Their inability to thwart me annoyed the Timbuktu mafia, but there was little they could do. Most of them sulked in Timbuktu, but Babaya had to stay in Araouane and represent the interests of his brother, the chief. He claimed to be the most enthusiastic of supporters of the project, but whenever problems with outside authorities arose or dissent brewed among the workers, he generally turned out to be the person responsible. He tried to convey the impression that he was my very best friend, and he gave me an open invitation to eat at his house as often as I wished. The fact remained that Babaya had lost the position of unchallenged power he used to enjoy, and even if he approved of the garden, he couldn't help trying to undermine it in small ways.

Sometimes Babaya would come down to the project and carry water. He wouldn't work all day like the rest of us, but I was glad that he made an effort even for an hour or so. He'd spend most of his time talking with the other workers, however, and I always waited to see what problem would pop up once Babaya had gone back home to his women and his tea.

One day when it was too cloudy for our solar-powered water pump to work, Babaya lent us his rope to pull water from the well. By the end of the day the rope had disappeared.

"See, they steal," Babaya gloated.

A few days later one of the kids told me he'd seen the rope in Babaya's own house.

Once he tried to sell me thirty camels (his entire herd, he claimed) for the equivalent of $14,000, but I refused to be suckered. I knew he owned many more than thirty beasts, and that the ones he'd give me would probably collapse from old age as soon as the money changed hands. Another time, Babaya and I both happened to be down in Timbuktu at the same time. He asked me to transport some supplies for his family back to Araouane in my truck.

"Of course, Babaya," I said, "no problem."

He knew my scale of fees (one goat or five chickens for every passenger or 100 kilograms of luggage), and had eight sacks deposited in front of my door.

"Fifty kilos each," he said. "You can take four goats from my herd in Araouane."

"Babaya," I protested, "these sacks must be more than fifty kilos each."

"No, no, I weighed them myself. Trust me."

When I reached Araouane, I weighed the load and found that he'd underpaid by one third. Brother of the chief or not, I decided, Babaya was not going to get away with it. Next time I went to Timbuktu, I told people that I would not have anything to do with Babaya anymore. I would not challenge his authority in strictly traditional matters, of course, but he was not to set foot on my project again.

The old cheat quickly heard about it and came around to my hotel to sort things out. I refused to see him. For the next few days he dogged my steps around town, until finally I shouted abuse at him in the middle of the big market.

"Liar!" I yelled. "Cheat! Thief!"

Babaya kept asking me to forgive him, but I would not.

One day he came with a man I didn't know, while I was loading the truck in the same marketplace.

"I hope you will listen to me," said the man, "even when I speak for Babaya."

"Okay," I said, "what do you have to say?"

"Babaya wants to tell you that he is very sorry. He admits he cheated you, and he knows now that this was wrong. He wants to live in peace with you, and will accept any punishment you deem fit."

There was a small crowd of people surrounding us. Whenever I went down to Timbuktu, people gathered around, fascinated by my customized truck.

"I accept the apology," I said, "but not for Babaya's sake. His kind of behavior makes it almost impossible to accomplish anything at all. I accept for the sake of his son Baba Boatna." The boy was one of the most promising young people in the village, and I did not want him to suffer for his father's tricks. "Babaya will have to pay double for the excess weight he put on the truck."

Babaya agreed, we shook hands, and he came back to Araouane with me.

I still didn't trust Babaya any farther than I could throw him—and he is quite heavy—but after that he worked every day in the project, pulling water from the well and carrying buckets, just like his servants used to do. Baba Boatna, moreover, cheerfully did any task that one asked of him, even when some of the blacks made sure that he ended up with most of the heavy lifting jobs.

Moreover, Bou-djema and Baba Boatna became insepara-
ble friends. I taught the two of them how to drive the Land-
Rover and perform routine maintainance work on the cars.
Before long they were driving almost daily to the banco
fields, competing with our ten donkeys to see who could
bring the most clay into town for construction projects.

Over time, Babaya abandoned his opposition. Despite his di-
minished stature in the village, he began to spend more of
his time in Araouane than in Timbuktu. On Fridays, though,
he continued to send somebody else to collect his rations.
The idea of accepting a handout from Boudj, Zain, or one of
the other blacks was too much for him. But he was becoming
genuinely civic-minded, and he tried to carve out a new political role for himself within the structure of the garden project. When I mentioned that we might rest on Fridays and
Saturdays but the trees could never take a day off, he organized a socially integrated group to do extra work on the
weekends. The squad included the Arabs Araouata, Ouaice,
and Abidine, and the Sorai Hammed Hella, Jambarka, and
old Baba Cambouse (when he wasn't too busy giving
Qur'anic instruction to the children).
    Babaya also took the initiative in solving another of our pe-
riodic problems, the difficulty of drawing water on days too
overcast for the solar pump to function. Even on bright days
the steadily growing influx of nomads strained our wells to
their limit. We couldn't turn these people away; not only
would it have violated the unwritten code of the desert, but
we relied on the dung from their camels for fuel.
    Babaya proposed that we dig a new well near the garden. I
was glad of his offer, but wondered what his angle might be.
Still, I bought a couple of old camels to pull the sand up from
the hole by means of a makeshift pulley system. Once we began to dig we found the enclosure of an old well which had
filled up with sand sometime in the distant past. Rather than
excavate an entirely new well, we decided to reactivate this
one.

Although the hole was still shallow, it was easy persuading people to work. All the children vied to get into the act, and it actually seemed fun. Day after day Babaya would oversee the work, leading the camels through their ever-lengthening circuit, pulling buckets of sand over our crude pulley system to the surface. He even let his son go down the hole, when it was already fairly deep. When the camels were worked to the bone from their incessant sand hauling, we cooked them up and ate them, then resumed work with another pair.

As the hole got deeper, volunteers to go down got more scarce. After thirteen days, there were none at all. Babaya sat gloating on the pile of sand we'd brought up.

"None of them will work," he said. "They are all scared." He gave me a smug grin.

I called everybody on the project together. "Who will go down?" I asked.

Silence.

"What is the problem?" I demanded.

Boudj's older brother Ahmed spoke up. "I was the last one down there," he said, "and I am too frightened to go back. It is completely dark in the hole, and very, very hot. The sand is wet so water must be near, but I could not see a thing."

"I know you're not accustomed to this kind of work," I said, "but your ancestors were. Araouane exists because your forefathers spent all their time digging and maintaining wells. Here is what I propose: Ahmed will go down for only two more hours, since he has been the bravest one so far, and after that Girage and I will take turns until we reach water."

I selected Girage for the nasty job because somehow that strong boy had ended up with all the cushy jobs in the project. He'd been given the post of custodian of the hotel, and when there were guests, he, along with a few other kids who had mastered a bit of French, earned 200 francs per visitor per day. This sum was the equivalent of less than a dollar, but in Araouane it represented a very nice wage.

He flatly refused. "I cannot go down," he said. "I get stiff even thinking about it."

He wouldn't be budged, not even when I threatened to take away his hotel job. So Ahmed agreed to work first, and I promised that when he came up for his midmorning *crème* break, I'd go down to finish the job.

Ahmed worked for a few hours, but by 10:30 he still hadn't come up. The rest of us started to get a bit worried. I called down to make sure he hadn't been injured, but he refused to come back up. When he eventually emerged, he was wearing a big grin. On his own initiative he'd completed the digging.

For lunch we cooked a huge pot of rice with vegetables from the garden in water from the new well, and we slaughtered and roasted the second pair of camels we'd been using to haul sand. We had a lot to celebrate: For the first time in many, many years, the town had a new source of water. And we were eating vegetables we'd grown ourselves, right here in Araouane.

During my third year in Araouane, Jean-Yves Pajot, the representative of the French industrialist who had donated the village's solar-powered water pump, came back for a visit. He was an avid jogger, and even though he was over sixty, he ran ten kilometers at least three days a week. On his previous visit, he'd challenged Mohammed Ali to a race and had been very impressed when the schoolteacher had left him in the dust. Mohammed Ali was a former 800-meter Malian champion, but he hadn't trained in a while, and he wasn't wearing any shoes.

Now Jean-Yves demanded a rematch. He'd brought some fancy track sneakers with him from France and had been psyching himself up for a long time. I proposed making the course the periphery of the garden, but Jean-Yves cannily refused: In the deep sand, he said, the desert champion would have an advantage. So we drove to a patch of land not far from town where the sand was hard-packed, and set up a pole at 500 meters from the car. Twenty laps between the car and the pole would be a ten-kilometer race, and since Mohammed Ali was not a distance runner, the contest would be fair.

All of the villagers, of course, were rooting for Mohammed Ali. After his last victory, they expected him to win easily.

"Even I could beat Jean-Yves," Babaya boasted.

"Okay," I said, "why don't you run too?"

"I have to work," he replied, "otherwise you'd say I was setting a bad example for the others."

"Sport is hard work too," I said. "It's healthy, and might stir up interest in physical fitness."

"All right," Babaya said, "I will run."

The next morning when I met up with Mohammed Ali and Jean-Yves to drive them to our new racetrack, Babaya was nowhere to be found. His wife, Kia, had no idea where he had gone. We left without him, but I had to turn back when I realized that I'd forgotten my camera. Back in town, I found that Babaya had mysteriously appeared.

"So, you didn't want to enter the race?" I asked.

"I wanted to," he answered, "but last night a few donkeys got loose and I had to go look for them."

"You're in luck," I said. "We haven't started yet, so you can still take part."

At the starting line Jean-Yves was decked out in all the latest running gear. Mohammed Ali wore a pair of old ski pants and a tattered sweatshirt of mine with a faded "USA" still visible on the back; Jean-Yves had brought an extra pair of sneakers for him, but Mohammed Ali opted to run barefoot instead. Babaya wore his usual flowing green jellaba, black turban, and billowing black pants. On his feet were plastic flip-flops.

"Does one have to run the whole distance without stopping?" he asked me.

"No, Babaya, you can walk from time to time."

"I have never run before."

"But you have walked with caravans, all the way to Timbuktu. Jean-Yves has never done anything like that. What's more, he is much older than you."

"That's right," Babaya said, with new confidence, "and he is no Arab!"

They were off. After the first lap Mohammed Ali was far ahead, with Jean-Yves second and Babaya well behind. He had already jettisoned his flip-flops, and during the next lap his turban came off as well. But the distance between him and the other two grew steadily greater.

As he rounded the post a third time, I yelled, "Babaya, take off the jellaba. It will be easier to run."

I knew he'd be embarrassed if the villagers saw him half-naked, but no one else was out near us. They all were at work. By the next lap he had shed his robe, and his sizable belly flopped over the drawstring of his pants. He was puffing and snorting, but his face retained a determined expression.

Off in the distance I saw a nomad with a couple of camels. I hoped for Babaya's sake that the man would not approach. Of course, he made a beeline for us. Anytime a Moor sees some activity in the desert, his curiosity compels him to investigate. Mohammed Ali ran by, then Jean-Yves. The nomad shook his head and looked at me as if to say, "What on earth are these people doing?"

Then he saw Babaya run up, huffing and puffing, dressed only in the skimpy pantaloons meant to be worn under his jellaba. The nomad was plainly baffled, but as Babaya came closer he nevertheless started the customary litany of greeting:

"*Salaam aleikum*—peace be with you."

"*Aleikum salaam*," Babaya puffed, not slowing down.

"May God protect you!"

"May God be with you!"

"No evil to you!" shouted the nomad. Some unintelligible sounds came from Babaya's direction, but he had already jogged out of range.

"I trust your health is in the hands of God!" the nomad

called after the running figure. "God be praised! God is great!" he added, receiving no response. He shot me one more confused look, shook his head, and retreated into the desert muttering, "God is great! God is almighty! God be praised!"

Mohammed Ali won the race, and Jean-Yves came in a distant second. We all waited quite a while for Babaya to finish, but finish he did.

EMILIE

My relationship with the women of Araouane was never an easy one. At first I had no way of dealing with them at all—the wall between the genders seemed too thick to be broken down by any *toubab*. But the greatest change that Araouane brought to my life resulted from this very difficulty. It brought me Emilie.

When I first visited Araouane with Dah's caravan, I had no idea how an outsider was supposed to behave around women. I knew that in many traditional Islamic societies, women avoided any contact whatsoever with men outside their families, so I decided to err on the side of caution. But the women of the desert did not always seem so cautious themselves.

One day during the trek from Timbuktu to Araouane, I was photographing the nomads who had joined our party. The men were so eager to have their pictures taken that I tentatively asked if it would be possible to photograph the women as well. With the laughing encouragement of the cameleers, a little boy took me by the hand and led me to the women's tent. The good-natured screeching and screaming that greeted my arrival was ear-splitting. All of the women, except Dah's wife Ouija, scrambled to cover their faces. Gradually the screams turned to giggles. Crazy Sidi's wife unveiled her face for a moment to speak, and whatever she said had the whole tent roaring with laughter. Then some of the women pushed a robed figure toward me, playfully pulled her veil askew. The girl quickly covered up again, but not until I'd

gotten a glimpse of her very attractive face. I couldn't understand what was being said, but it seemed that the woman
was unattached and all the wives were having a joke at our
expense.

Ouija made some room for me to sit next to her, and
poured me tea. I gave the camera to the little boy who'd
brought me there, and gestured for him to take a picture of
me with the nomad women. The women all started to sing
and to beat on buckets or teakettles because they didn't have
their tam-tam drums with them. The photo shows me surrounded by a bevy of females who I'd been told would run
from my very sight!

The day after Dah and I had arrived in Araouane, a pretty
young girl started hanging around my hut. Every time Dah
was gone and there was nobody else around, she would
mysteriously show up. She just sat there by the door, looking
awkwardly at the ground. We had no common language, so I
had no way of asking her why she had come. I wondered if
she was the town's "hospitality" for an honored guest, but I
didn't dare find out.

When I came back to Araouane to start the garden, I again
had a tough time figuring out the rules. After many weeks in
the desert, I naturally found myself thinking about women
often, but I did nothing more than think. It wasn't all that
hard to keep my eye from wandering—not only did I want to
avoid giving offense by staring, but the women who carried
water for the garden were all so thickly draped that I had no
idea what they really looked like.

After I'd been living in the village for a month or so, one of
the women began removing her veil every time she came
near me. Several workers at different times made it a point to
tell me that her name was Nana; until that point I hadn't
known the name of any woman in Araouane. What's more, I
was told in hushed tones, Nana had no husband.

As the villagers and I became more familiar with each
other, the women started to push Nana into me whenever I
passed. They'd stand there afterward giggling and saying

things I didn't understand. Eventually Nana took to lying in the sand, striking suggestive poses, while the other women pointed at her and at me.

I lay awake many a night on my millet-sack bed thinking about Nana, but I couldn't think up any way of initiating a relationship with her. I spoke barely a word of her language, and she spoke even less of mine. Moreover, the daily village routine provided no opportunities for the two of us to be alone. Every time I saw her, she was in the company of several other women or family members. Even if I somehow managed to sort out the logistical difficulties, there was a more important obstacle: I couldn't imagine marrying a complete stranger (and one from a wholly alien culture), and I feared that a casual fling could destroy her social standing.

I had resigned myself to celibacy during my time in the desert, and regarded sexual tension primarily as an inconvenience. By the time Nana came to my notice, I'd grown accustomed to dealing with it by, shall we say, taking the matter quickly in hand. But once this beautiful young woman began invading my dreams, my peace of mind went right out the window. Go for it, I'd say to myself. No, I'd answer, the whole village will be talking about it the next day. At times I feared that if, by some miracle, Nana magically appeared on the millet sack beside me, I'd be as nervous as a teenager and forget what to do.

Eventually I was cured of my infatuation with Nana. I was working near the construction site of the new schoolhouse when I heard a loud thump. I turned to see a cloud of dust puffing out from the doorway. The schoolhouse roof had collapsed because we hadn't let the banco walls dry sufficiently before putting on the ceiling. Three boys who had been working inside were trapped under the rubble.

The village broke out in total pandemonium. All of the townspeople and most of the Timbuktu workers fell into a frenzy. They hacked at the debris with hatchets, axes, and shovels, chopping so madly that they would surely have cut the moaning boys underneath to pieces if they'd reached them. I tried to restore some calm, but it was a scene of complete hysteria. Finally I had to knock one of the Timbuktu men out and violently shove the others out of the way to impose some order and pull the boys to safety.

They were bruised and bloodied, and of course badly shaken, but not seriously injured. Mohammed Ali, Boudjema, and old Baba Cambouse were the only villagers collected enough to lend a hand. First I took care of the wounded boys (mostly with painkillers and sedatives), then I turned my attention to the onlookers. They were such a blubbering, shaking, moaning crowd that it looked as if a shock grenade had been tossed in their midst. I used up all the tranquilizers and sleeping pills I'd brought with me from Switzerland.

The worst of the bunch was Nana. Her brother was one of the injured boys, and when she saw him, she melted into a trembling mass of jelly. She tore off part of her robe and lay writhing in the ground, her mouth foaming with saliva. In her hysteria she clawed at her beautiful breasts until they were bloody. She was screaming and kicking so hard I couldn't get a sleeping pill into her mouth, and had to dissolve it in water to pour down her throat while I held her nose. From that day onward, I never again dreamed of Nana.

When it became clear to the villagers that Nana and I would not become a couple, Garba's wife started showing up every morning at my house. Sometimes I woke to find her seated beside my bed. She was very pretty, but this was too much. Taking up with another man's wife would certainly have ended my welcome in Araouane. When she started following me out to the dunes to watch me take my morning piss, I had to forbid her to come to my house anymore.

All these missed opportunities seemed to have given the villagers the impression that I wasn't interested in women at all. Once an important Timbuktu man took me aside, wearing an earnest expression on his face.

"Aebi," he said, "would you like to sleep with the boy Boudjema?"

The sexual norms and customs of Araouane were generally shrouded from me. Only gradually did I learn what was accepted and what wasn't, and even then I only got part of the picture. As an outsider, and as a man, there was much that I would never truly know.

I had often wondered if the women of Araouane were subject to the mainly Muslim custom of clitorectomy.

One day Boudj brought me a little rusty blade to sharpen. The knife was a type I had never seen in the area, and I asked him what it was used for. He told me that his grandmother was to make "little cuts" on his half-sister. I asked him if he thought it would be a problem for me to photograph the procedure. He said he thought it would be okay and promised to come and get me when it was being done.

A bit later he reappeared to say that the ceremony had been put off because strong winds had raised too much dust. A few days later I had to go to Timbuktu, and when I came back, I was told that it had been performed during my absence. I asked Boudj what exactly was done. He told me that his grandmother made little cuts around the waists—he pointed a little above his hip joints—of all the little girls to produce their customary scars. But I didn't notice any such scars on the hips of the often scantily clad little girls.

When I asked Mohammed Ali about clitorectomy, he seemed very uncomfortable, as he usually was when any sexual reference was made, and strongly maintained that no such practices took place in the village, at least to his knowledge. "That's mostly a Bamabara practice," he said.

I tried subsequently to find out more about the subject by bringing it up, as casually as possible, but no one appeared to know what I was talking about. Although sometimes I had the strong suspicion that the subject was simply off-limits to men.

Despite the Timbuktu man's suggestion, I never noticed any homosexual behavior in the village. Men would often hold hands, cuddle, and hug each other, but as in many places throughout Africa, this behavior was a common way of showing friendship. I likewise saw little evidence of heterosexual unions outside marriage. Adultery seemed almost impossible, since a village woman typically spent virtually every moment in the company of either her family or other women.

It was not at all unusual, however, for a wealthy man (almost always a white Moor) to have several wives, some black and some white. In such cases, the children all called one another "brother" and "sister" regardless of the respective races and statuses of their mothers. They made a distinction between full siblings and half siblings, but did not split into separate families based on skin color.

Marriages were generally made by family arrangement rather than for love. The first wedding I saw was that of a girl named Agida, during my first year. Boudj passed the invitation along to me, and when I showed up at the house, blankets were set out and several women were banging on tam-tam drums. Each family brought a dish of food to share, and the mood was festive. When I asked Boudj who the groom was, he pointed out a young fellow I didn't recognize.

"Has he known Agida long?" I asked.

"No," Boudj answered, "he arrived in town only yesterday. He has had business dealings with the girl's father, so he asked for her as a bride. He gave the father much salt."

The marabouts recited verses from the Qur'an inside the father's house, and everyone sat down to the meal. Afterward the women made up a luxurious bed in an empty house nearby, so that the bride and groom could spend their wedding night in privacy.

The husband eventually went back to the mines at Taoudenni, and a year later he sent Agida's father a note saying he no longer wanted the girl as his wife. Mohammed Ali told me that several weeks went by before the father even bothered to tell Agida that she was divorced.

Another wedding I saw was more elaborate, but less complete: There wasn't any bride present. Dah, a son of the deceased village chief Mohammed Sultan, was getting married to a daughter of my guide at the time, Mokhtar Moulay. The groom butchered a camel and several goats to provide a feast for the town. For three days the assembled company ate and drank tea and talked well into the night. On the third day the festival ended with a wild camel race. Nobody seemed to mind that the bride was only a little girl who wouldn't even meet her husband for years. She was at the time only about eight years old, too young to take part in the festivities at her own wedding.

The villagers' complicated mores and taboos around sexual behavior mystified me. Even Mohammed Ali couldn't understand many of the local practices. For example, husbands often went out of their way to leave me alone with their wives when I was acting in the capacity of a doctor. When the wealthy Moor Araouata brought his beautiful wife Silka to me for treatment of an earache, not only did he leave the room, but he insisted that Boudj leave as well. I later learned it would have been insulting for him to remain, as it would have implied that he didn't fully trust me.

Yet the men went to extravagant lengths to avoid even being in the presence of their brothers' wives. Araouata's brother Ouaice, for instance, would not set eyes on Silka, let alone speak to her, nor would Araouata set eyes on Ouaice's wife. When Araouata wanted to visit his brother's house next door, he would go to the mosque and wait until someone passed by. He would tell the man to run to Ouaice's house and say that Araouata planned to visit, so that Ouaice's wife would have time to leave before he arrived. Often she would go and have tea with Silka, so Araouata would have to repeat the process in reverse to go home.

With such a complicated etiquette in place, it is little wonder that I usually felt in danger of breaking some obscure rule governing relations between the sexes.

I had no missionary desire to tamper with local mores, but I did want to give women the opportunity to work on the project beyond their traditional chore of hauling water. As a man, however, I had little opportunity of talking freely with them. At first, my only female recruits were a young girl and an old woman.

The girl was Agida's sister Amma, one of my favorite children in the town. Unlike the other women, she seemed to have no taboos about what sort of work she could do, who she could talk to, how she could dress. She was a genuine tomboy of the desert. She would gladly carry banco for making bricks and perform any other tasks usually reserved for boys or men. Women never took part in the butchering of camels or goats, but whenever there was an animal to be slaughtered, Amma was always in the thick of it, pulling and ripping on the bloody pieces of meat to make it easier for the men to cut with their dull knifes. It was obvious that the adults of the village found her behavior scandalous, but Amma didn't care. I found out later that her father, Hababou (an obnoxious loudmouth), had told her to work hard at any job I assigned, so that if the garden was a success, he would have a stake in it from the start. Amma didn't care about the politics of it; she just loved the chance to work.

I almost thought we had lost Amma once. Mokhtar Moulay, the guide who accompanied me on excursions to Timbuktu, woke me from a deep sleep.

"We have to get the Land-Rover," he said. "Amma went to the desert to search for camel dung and never came back."

For several hours we drove in wide circles around the village. Our task was beginning to seem hopeless when Moulay noticed a signal fire on the highest dune in town.

"That's the sign for us to come back," he said.

The villagers had found Amma, safe and sound. She had returned hours before, so exhausted from her work that she'd fallen asleep behind a big pile of ropes.

Typical of Araouane's confusing (for me) social customs, I heard about Amma's marriage before she did. One afternoon Boudj told me that she was going to be wed that evening to his elder half-brother, but that I shouldn't tell her because it would be "a surprise."

• • •

My other female recruit was Tata, a spindly, frail lady so ancient that she no longer felt bound by the usual conventions.
Every time I returned from a trip to Timbuktu, she hugged
me fervently, as if she'd never expected to see me again. She
always dressed in rags, and never wore a veil. She wore a
perpetual grin that gave her the air of a favorite nutty aunt.

In the desert surrounding the town, I'd often see shards of
crockery poking out of the sand. Some of the villagers still
used unglazed ceramic jugs to store their water, but most
used ugly iron bowls imported from China.

"Why do you use these metal containers?" I asked Bou-
djema. "The pottery vessels are much prettier, keep the water
cooler, and you don't have to buy them in the city."

"Nobody in the village knows how to make them anymore."

"You must be kidding," I said. "Didn't people used to make
them here?"

"Yes, a long time ago," he replied. "But today only old Tata
knows, and she hasn't done it since she was a little girl."

"Come on," I said. "Let's talk to Tata."

We went to the old woman's house, but when she heard
my request, her grin turned to a scowl. "I don't remember
how to do it," she said. "Nobody has made a pot here for a
very long time."

"That is very bad, Tata. As the smartest of the older wo-
men, you had a duty to pass your knowledge down to the
young girls. Without your help, the tradition will be lost.
What if the townspeople cannot get iron bowls anymore?
What will the little children eat from? If the knowledge dies,
God will be very angry with you."

"Aebi, don't treat me like that," she said. "You know I am
an innocent old woman."

"Not so old, Tata. At Gomeini's wedding you wore me out
dancing."

"That's not fair. Of course I was full of energy then—I was
excited because Gomeini is the son of my sister."

"Don't be difficult," I said. "Tomorrow we'll go and get the clay. Do you remember where to find it?"

"It's too far in the desert, a long way north of here. An old woman like me would never find it."

I sent Bou-djema to ask some of the old men if they remembered where the clay deposit was.

"Tomorrow morning we'll come and get you," I told Tata sternly. "And I don't want to hear any excuses. Some of the old men will come with us, and we will certainly find the clay."

The following morning we drove north, with old Bella, old Hamma, Tata, and my two helpers, Bou-djema and Baba Boatna, all crammed into the Land-Rover. None of the old people had ever been in a car before, and the longer we drove, the more excited they got.

After about half an hour, Tata told me to stop the car. She was sure the clay was somewhere nearby, and the others agreed. The three skeletal figures spread out over the desert, their tattered robes flapping in the wind. Bou-djema, Baba-Boatna, and I drove onto the highest dune to make sure we kept all of them in sight. We could not help in the search because we had no idea what the clay deposits looked like.

They all came back empty handed. I wondered whether they had wanted to stop because of the clay or because they were getting carsick.

"Do not worry," Boudj told me on the way home, "Chou Monsoor knows where the clay is."

Chou Monsoor was an old man, the father of Nana and young Mokhtar, who lived in a one-room hut on the northern edge of town. He hardly ever mingled with the other villagers and often disappeared for a few days at a time. Mohammed Ali thought that he was helping nomads with odds and ends. The only time I had dealings with him was when he came to me for help with an ugly infection from a scorpion sting on his belly.

"Why did you bring these old folks instead of him?" I asked Boudj.

"Chou said he would not get into a car."

"We can go find the clay with him on foot, then go back to fetch it later with the car. What's the problem?"

"Tata has to see the clay before we bring it back," Boudj said. "There is good and bad clay, and she cannot walk that far."

It seemed the whole village was in a conspiracy to avoid getting the clay. But when Boudj explained our dilemma to Bella, the old man flashed a toothless, wicked smile and promised to get Chou into the car.

He came through. I have no idea what he said, but Chou looked petrified as he got into the Land-Rover. And only 8 kilometers north of the village, he found the clay. Tata examined it and declared it to be the right kind. I took out a shovel, a pick and a couple of buckets to start digging for the stuff.

"No! No!" Tata shrieked.

By this time I was quite annoyed. We'd spent the whole day in search of the clay, and now that we'd found the stuff, she didn't want to take any of it. I ignored her pleas and started to dig. Tata threw herself against me with surprising strength.

"God will be very angry with you," she cried, "if you take clay without giving Him an offering of thanks. Tomorrow I shall bring one, and then we can take as much clay as we want."

I didn't want to get into an argument with either Tata or Allah, so once more we drove back to the village without clay.

The next day, with a whole bunch of kids in tow, Tata and I returned to the desert. Before the children and I started to dig the clay up and shovel it into buckets, Tata laid out little balls of millet paste. She looked them over and shook her head.

"I thought that this would be enough," she said, "but when I see all that beautiful clay, I know Allah wants more than just the millet. To be safe, we should bleed a chicken here as well."

"Tata, we do not have a chicken here." I was really getting exasperated.

"We have to get one in the village."

"No, no, that's quite enough. Now, we get clay."

"I would be scared to make a vessel with clay that God might think had been stolen from Him," she said. "I just could not do it."

I left Tata and the children in the desert, drove back to the village, and grabbed the first chicken I saw. It belonged to Amma's father. I yelled to him that I'd pay for it in the evening, drove back to the clay pit, and threw the chicken at Tata's feet.

The rites took only a few moments. Tata took a little blade out of her necklace, cut the chicken's neck, let the bird flap around until it died, and beamed at me.

"All right, Aebi," she said happily, "now we can take all the clay we want."

Tata, it turned out, had no artistic skill at all. Her pots were misshapen lumps, which was probably why nobody had wanted her to make ceramic vessels for them. But at my insistence most of the village women watched her preparing the clay, shaping, drying, and firing the pots. God willing, at least one of the other women would develop better aesthetic sense. At any rate, the traditional techniques had been preserved.

During the first two years, this was about the extent of my interaction with the women of the village. Mohammed Ali had no more success talking with the women than I did, so it seemed the difficulty stemmed from my being a man rather than from my being a Westerner. A few women were even heard to mention that they'd like to learn French, but that it was unthinkable for them to enter a classroom with a male teacher.

Even so, the project brought modest improvement to their lives. Previously the women had spent most of their day alone or with their children, leaving their homes only to draw water or collect camel dung for their cooking fires. The garden let them develop a sense of community. At first the bucket brigade had worked in almost total silence, barely exchanging words even when they took turns filling their pails. Over time the women got increasingly sociable, and soon they were joking and conversing all through the work day. When the laughter, shouting, and gossip reached schoolgirl pitch, I sometimes longed for the glum old days, but the women certainly seemed to enjoy their daily routine more than in the past.

That didn't seem like enough, though, and it frustrated me that so large a pool of labor for the project was barely being tapped. But I knew that it would take more than Mohammed Ali or I could provide to make any significant advances toward social integration. It would take a living example.

When I returned to America after my first year in the Sahara, I spread the word among my friends and acquaintances that I would like to take a female companion back with me to Africa. She would have to speak French, know a thing or two about gardening and crafts, be able to teach school and drive a Land-Rover in the desert. It was vital, of course, that she be somebody with whom I could get along—we'd be too close together and too far away from anyone else to avoid each other if we didn't hit it off well, and it wouldn't have set a good example for the villagers if we were always getting into fights. Most of all, she would need a sense of adventure and a willingness to undergo physical hardship without complaint. Privately, I also hoped she'd be beautiful.

As it turned out, there were quite a few women interested in this mad undertaking. I'd thought I might have to beg or cajole some acquaintance into coming, but I found myself in the enviable position of having to interview applicants. Unfortunately, nobody suitable turned out to be willing to give up her job, apartment, or family for the year or so required at such short notice, so that second year I went back to Araouane accompanied only by Fritz.

During my time in New York, however, I fell into a very nice relationship with a photographer I'd met. Emilie had grown up on Long Island, and set out at twenty to go to school and work as a photographer in California. After a few years she returned to the Big Apple, "because that's where everything happens."

One day I brought some rolls of film into the photo lab where she worked and soon I found I had an almost daily need for all sorts of camera equipment and photographic services. When I came back to New York after my second year in Araouane, Emilie decided to return with me in the fall. She understood my plans for the village women immediately, and took intensive courses in French and gardening to prepare herself.

My own motives were more than simply altruistic. I was finding the desert more than a bit lonely, and looked forward to having a female associate there who was more than a coworker. I rather liked the prospect of spending every day for six or seven months in Emilie's company, and she didn't mind either.

We started off in Switzerland and drove across the Sahara, traveling with a camera crew at work on a documentary about the project. Emilie was a bit disappointed that instead of a romantic solo trip through the desert we had a two-car convoy and inquisitive filmmakers who were always pointing their lenses at us. But the lack of privacy turned out to be a blessing: The Tuareg uprising forced us on a roundabout 1,400-kilometer drive through the wasteland, and I wouldn't have dared make the voyage without a backup automobile.

There were plenty of opportunities for panic, but Emilie never lost her cool, even through the worst part of the trip— an interminable stretch of desert with no wells, no nomads, no tracks, nothing but uncharted dunes and rocks. At one point it took us two and a half days to cover only 8 miles. Often we had no choice but to shovel our way through a mountain of sand, keeping a nervous eye on our supplies of water and fuel. We rebuilt a wheel that had been destroyed by hard wear, and even had to abandon a trailer. But Emilie didn't let the hardship bring her spirits down. Long afterward, Bill, the cameraman of the film crew, told me of the reactions he received when he showed friends the shots of our trek.

"For God's sake," they said, "who is that woman wearing pearls in the middle of the Sahara?"

"You should see her dig out a Landcruiser," Bill replied.

Emilie took readily to the desert, to Araouane, and to all the hard work. There was nothing she wouldn't try or couldn't tolerate, and she came to love the desert dwellers as much as I did.

Not long after Emilie had arrived in the village, Bou-djema gave me his opinion: "You found the best woman in America to bring here."

At the first village meeting, I introduced Emilie as my wife. It was not really a deception, since by local custom any man and woman living in the same house are considered married. If they feel like it and can afford it, the family throws a feast. Divorce is equally simple, at least for a man—he just tells his wife that he doesn't want her anymore, and that's the end of it. For a woman it is more difficult, but a woman would have to be truly miserable in a marriage to brave the misery of life as a divorcee in Islamic society.

A few days later, some of the women and children caught a glimpse of us embracing and thought it was the funniest thing they'd ever seen. Couples in the village never displayed physical affection in public. The women took to asking Emilie and me to hug or kiss for their entertainment, then they'd giggle, punch, and slap at each other when we obliged. They never got bored watching us embrace. We didn't mind, but we felt we had to set some parameters.

The women kept urging us on and would probably have been quite happy to watch us make love right before their eyes. At first they started greeting Emilie and me with lewd gestures. The very vegetables in the garden took on a whole new meaning: Every time one of the women harvested a large zucchini, a turnip, a carrot, or an unusually elongated egg-plant, the whole female work force would erupt in laughter.

"Emilie!" they'd yell, pointing at the vegetable, "Emilie—Aebi!"

The women warmed to Emilie almost instantly. Her communication with them was initially limited to gestures, giggles, and hugs, but that didn't seem to matter. She helped them set up their own garden plots, organized all-female work details to make seed beds from rotting camel dung and build protective walls out of banco. It made me very happy to see the women hurrying around with bricks and shovels, striving to make their plots better than those their husbands had constructed.

When the men saw how ably their wives could garden, they started ordering the women to take over their own chores. Soon it seemed as if the women, who had until recently been barred from doing any work other than carrying water, would be doing all the work on the project.

Emilie put an immediate stop to this. She pushed the women to stand up to their husbands, and scolded men who tried to make their wives do their own work. Once she discovered that the plot which had been assigned to Yumma, who'd become a close friend, happened to be in front of Babaya's house. It was a rule of the project that each family would own the portion of garden directly in front of its own home, but it looked as if Yumma would be preparing a plot that would eventually go to Babaya and his son Baba Boatna. Babaya had delegated all gardening matters to his boy, so we called Baba Boatna and Yumma together to find out what was going on.

"I don't want Yumma to put in so much work on her garden," Emilie said, "just to find out later that it's all going to be yours, Baba."

An Arab aristocrat, Baba Boatna, about the same age as Boudj, was one of the more hard-working and open-minded children in town. He stood up for his rights as an equal, but

I'd never seen him claim any special privilege on the strength of his position.

"Baba," I asked, "does this mean that Yumma is making a garden for you?"

"No, not at all," he said. "There was a part of my garden which I had not yet planted, so when the women got their plots I gave this bit to Yumma. I plan to make a bigger garden outside the wall, so I didn't think we'd need this smaller section. This way Yumma gets a bigger plot, and the land does not go to waste."

Emilie and I were overjoyed to see the son of the most influential Arab in the village sharing chores with a black woman. One of them would gather manure for both, and they'd take turns watering each others' gardens.

Once I came back to our house to see a nomad scurrying away as fast as he could. He tried to run toward the well where he'd left his camels, but his legs were entangled in such a mess of bandages that he looked a bit like a hobbled camel himself.

He had come to our house looking for Mohammed Ali or me, because he had big gashes in his feet that needed medical attention. Since we weren't there, Emilie and Boudj had decided to take care of him. They applied antiseptic cream and were bandaging the nomad's feet when their patient turned his head to get a good look at his doctor. He clearly had been uncomfortable at being ministered to by a white foreigner, but when he realized that the *toubab* in the shirt and pants busy tending to his toes was a *woman,* the poor fellow dashed out of the hut shrieking.

Another time some nomads came to the house with an old camel to sell. Again, I wasn't there, so Emilie decided to act in my place. She tried to buy it from them, but the nomads hurriedly took the camel and left. They couldn't imagine anything so preposterous as doing business with a woman. Only a few days later, however, another nomad came with a nanny goat and kid for sale, and Emilie struck a hard bargain. She

ended up paying about half of what I would have—the nomad, it seemed, had been so intimidated by the novelty of
haggling with a female that he agreed to a price far less than
market value.

Kia, the wife of the village boss Babaya, was a heavyset Arab
woman of such high social standing that she would rarely
deign to leave the house. She never associated freely with the
other women of the town, and since Babaya spent much of
his time in Timbuktu with his other two wives and families,
Kia usually stayed home alone. She had no work to do, no
friends to gossip with, and since she was illiterate (like all the
local women), she couldn't even spend her time reading. I
often wondered just how she occupied herself.

Apparently, she found her life of aristocratic leisure some-
what dull. During my second year's break in New York, she
started inviting Mohammed Ali over for dinner. She always
had Boudj, Girage, and her own son Baba Boatna there as
chaperones, but it was still completely unheard of for a re-
spectable matron to serve meals to a single man when her
husband wasn't around. Mohammed Ali later told me that Kia
showed a great curiosity about the garden project, local
events, and even the peoples and nations beyond the desert.
She had only left Araouane once, to visit Timbuktu with
Babaya in my Land-Rover, and had spent virtually all the
hours of her life inside the same little house, but her mind
had roamed far.

Over dinner and tea she usually asked Mohammed Ali to
tell her about the wider world. She asked him about the lives
of women in other parts of Mali, about the position of
women in other nations. She questioned him for hours on
end; sometimes he knew the answers, and sometimes he
didn't. For several weeks Mohammed Ali and the three children spent nearly every evening in Kia's house.

As soon as Emilie arrived, Kia invited her over for tea, with
Boudj serving as interpreter. Although they couldn't communicate very well, they quickly became friends, and to see

them hugging and slapping each other's backs you'd think they'd been companions from childhood. As their friendship grew, Emilie began to coax Kia out of the house.

Even the walk down to the garden was very hard on the heavy Arab woman. All those years sitting around inside hadn't given her any opportunity to exercise her leg muscles. Emilie invited her over for tea at our place, and by the time she arrived, her feet were so swollen that she could hardly stand. She looked like a patient with gout, and one of the village girls had to massage her feet while she sat and rested.

Emilie and I were glad she'd come, but we didn't want to treat her any differently from the way we would the other women. When it was time for her to leave, we thought of driving her back home in the car. But we didn't want the other townspeople to see us chauffeuring Babaya's wife around as if she were royalty, so Kia walked all the way home. In fact, she never demanded special treatment. Eventually she started coming over to our house when the other village women were there, to sit on the floor with them and share tea, laughing and gossiping like teenagers at a slumber party. Her physical condition kept Kia from working in the garden, but when the other women started a quilting bee, she joined right in.

When traveling from Timbuktu to Araouane, I always took along a navigator, so that I could concentrate on the desert rather than the course. First I used Sidi Mohammed, a former nomad who had lived in Timbuktu since his camels died; with him as a guide, I could never travel in a straight line— we zigzagged from one nomad camp to another. In Timbuktu he couldn't get the camel milk and other food he was accustomed to, so every time he got out to the desert he made certain to fill up on as much nomad fare as he could. He'd beg shamelessly, and eat so much that I often wondered whether our hosts had any food left for themselves. Eventually I grew annoyed enough to fire Sidi Mohammed in favor of Moulay, an Araouane man who had once worked as a guide for an oil exploration company in another part of the desert.

With Emilie's arrival, I no longer needed to hire a guide. Armed only with map and compass, she had already navigated us through the empty Tanezrouft, from Reggane directly to Araouane. After driving 1,200 kilometers through a stretch of desert with virtually no points of reference, the Timbuktu–Araouane run seemed like a breeze. When the villagers and the nomads had found out that Emilie had been navigator for this voyage, their awe was boundless. A desert guide, after all, is one of the most respected figures in any Saharan community.

One time a friend of Emilie's came to Araouane as a tourist. She had traveled from Timbuktu to the village with two con artists who preyed on foreigners around the Azalai hotel. Her trip had been horrible, since these two city dwellers knew almost nothing about life in the desert. Emilie decided that she herself would arrange the camel trek back to Timbuktu, through people we knew personally, to make sure her friend got a better feel for the desert.

We asked Araouata to provide us with four camels and somebody who could go with the women. With more tourists passing through our new hotel, we hoped that the villagers would soon be able to make money by taking visitors on desert safaris by camel. Emilie and her friend could be the first to test-market this new service.

The whole town got excited over the trip, with everybody making suggestions and offering advice. We loaded up the packs with plenty of food, water, and dates plus a supply of peanuts to distribute to the nomad kids who would surely materialize out of nowhere once the desert rumor mill reported on Emilie's caravan. Baba Hanta was chosen as guide and camel handler; he, Babaya, Araouata, and Hababou were the only men in the village who had ever made the trip to Timbuktu, and the other three were too busy with other caravans to leave. The women all laughed hysterically when they heard that Baba Hanta would be responsible for the two *toubabs*.

When the camels were loaded, the whole population gathered to accompany them out into the desert. Everybody was nervous. Araouata could barely talk. At the first range of dunes, he said a long travel prayer. The rest of us turned to go back to town, but Araouata kept going with the caravan.

"He wants to see them safely to the next ridge," Boudj told me.

We waited and watched. Araouata marched alongside the camels for a while, then began to march more slowly, occasionally looking back to see the village far in the distance. He stopped and started singing a wailing prayer, then he hurried back to us, afraid that he might not find the village again by himself.

Three days later Mohammed Ali arrived in Araouane by caravan. He had returned from his vacation in Timbuktu and was bringing food to the village—we had gone back to using camel caravans to haul supplies whenever possible, so that the villagers wouldn't come to depend on my truck.

"Did you see Emilie on the way?" I asked casually.

"Where?"

"Her caravan should have crossed yours. They are on their way to Timbuktu."

"Who is their guide?"

"Baba Hanta."

"What?" he cried, clearly shocked. "Baba Hanta! That man couldn't even find the village again if he went over a dune for a piss!" It seemed that this was common knowledge, but none of the villagers wanted to say anything bad about a fellow townsman.

Emilie had a compass, so I wasn't particularly worried. It turned out that a few days after leaving Araouane, Baba Hanta fell off his camel, landed squarely on his head, and was not quite rational for the rest of the voyage. Emile took over the navigation and guided them all to Timbuktu without a problem.

Emilie's presence helped the villagers accept me as a full-fledged member of the community. Every man should have a woman, the local sentiment went, so until she came people assumed I would run off and leave them.

In the spring of the year of Emilie, Dah came to Araouane for the first time since our initial voyage. It was March, and the garden was brimming with fresh vegetables. We hugged and laughed and hugged again. Out in the desert Dah had heard a great deal about Emilie and was very eager to meet her. He was spic and span, and his *boubou* looked as if it had just come from the dry cleaner's.

Emilie strongly impressed the nomad with her easy acceptance of desert living. Boudj served us a salad (even though Dah had once told me that tomatoes could be deadly), and Emilie nonchalantly picked a piece of goat shit out of the bowl.

"This doesn't belong in the salad," she observed, as she threw the lump away and continued to eat.

Once, when Emilie had gone to Timbuktu for supplies, some of her friends came over for a visit. They must have thought it was their duty to cheer me up now that my "wife" was away.

Ouaija and Yumma were cooking a large pot of rice with vegetables when Marijama joined us, dressed only in a skirt and a tank top. She'd found the outfit in a pile of cast-off clothing we'd brought with us from Switzerland. Ouaija and Yumma laughed uncontrollably and pushed Marijama into me. I asked her jokingly if she had become French. The poor girl was so embarrassed that she lay on the ground and tried to bury her head in the sand like an ostrich.

It was all in good fun, and showed how comfortable the village women had become around Emilie and me, but I didn't want to encourage the women to flout all the traditions. After all, we didn't want people from other villages laughing at Araouane or saying that the town had gone mad. I tried to have Mohammed Ali talk to the village women, to tell them that some of their new behavior would be unacceptable in the outside world. But the teacher, himself the product of a restrictive society, was too shy to broach the subject with them.

The men of the village were amazed at the transformation in their wives and daughters, but they didn't seem to mind. Emilie's presence made it clear that my intentions toward their women were honorable. When Emilie got back from Timbuktu, we decided to hold a race. All villagers were urged to participate, with men running ten kilometers and women only five. The first prize for men was half a goat with the skin, the innards, and head, and for the women the other—clean—half of the animal. Second prize was a generous supply of tea and sugar. Everyone who finished the race got 1 kilogram of flour.

Eight women and about twenty men entered the race. Emilie jokingly asked Djadja, the prettiest of the village women, if she would like to borrow shorts for the race. Djadja nodded enthusiastically, and was only dissuaded by the taunting of her less liberated friends. In the end all eight women ran in their flowing robes, and every one of them finished.

For the fifteen years since my divorce I had vowed, guaranteed, sworn, professed, assured, yelled, groaned, sung, and yodeled that under no circumstances, come hell or high water, would I ever get married again.

But Emilie had made life in the desert magical. Whereas previously I had struggled to communicate with the villagers or made eloquent speeches to the wind, now I could have long conversations with a kindred soul. Whatever little thing had happened during the day that made me burst with joy or seethe with frustration, I now had somebody with whom to share it.

Sometime in February of 1991 I started to develop serious throat problems. It always felt like I had something in the back of the throat, some sort of blockage that was making me choke. I had a very hard time merely breathing. It felt as if something was trying to work its way out, and my face would turn red, my skin would go clammy with the effort of preventing it. I developed chest pains and lost my appetite. I never thought I'd get sick in Araouane, yet here I was, miserably tossing about, gasping and panting for air.

Finally, late one night, I couldn't take it anymore. I woke Emilie up.

"Do you want to marry me?" I gasped.

The constriction vanished. I could breathe again.

We went to Manhattan's City Hall in May after my third year in Araouane, and returned to Africa in October as legitimate husband and wife.

L'HOMME DU DESERT

In April of 1990, my second year in Araouane, I flew back to America for seven months. That fall, like every other year, I went to my brother's house in Switzerland to prepare for my return to the village. I outfitted a new truck, and again contacted doctors and pharmaceutical companies for medical shipments and got relatives in Switzerland collecting crates of used clothing so that my friends in Africa would have something to wear on cold winter mornings.

When I returned to Araouane in October, Boudj ran up to me with big news: for the first time in forty-two years, it had rained. "There was so much water," he said, "that it reached up to my knees!" Mohammed Ali had set out a jar to measure the rainfall and calculated it to be 4.1 millimeters—not up to anybody's knees, not even the ankle, but far more than the desert had been blessed with in generations. "It is due to your presence here, Aebi," some of the villagers told me earnestly. "You come, and suddenly it rains! It is a clear sign that God is pleased with our hard work, and pleased with you for leading us."

The rain was a mere coincidence, of course, but my impact on Araouane was clear enough: A town that had been on the verge of starvation was thriving. Less clear, but no less striking, was the impact that Araouane had had upon *me*.

As I have mentioned, I more or less stumbled into Araouane and my project there. Like any youth, I had dreamed of adventure and exploration, and I had a few wild years on the road in my late teens and early twenties. But for most of my adult life I had put aside my dreams for the mundane work of raising children. I'd had a very messy divorce, which left me flat broke and with the sole custody of four young children, which led to two decades of PTA meetings, dentist appointments, chicken pox, and school principals lecturing me on how to improve my "parenting skills" so the children wouldn't be suspended so often. All this while I had to earn a living.

Through it all I longed for faraway places and great discoveries. I leafed through the atlas looking for "white areas," unexplored territories waiting to be mapped, remote places like those that drew Livingston, Humboldt, Baker, Burton, Speke, Caillié, and Barth. I fretted that maybe there were no such places left anymore, out of the reach of telephone, telex, or satellite dish. I envied Magellan, Marco Polo, and Amundsen, Scott, Vespucci, and Leif Ericson. But all I could do was read books about the great explorers of the past and bemoan the fact that I had arrived too late.

I grabbed at straws. I took up wreck diving in the New York harbor. Here I could have the excitement and danger of exploration and still get home in time to fix supper for the kids. My diving buddies and I would grope through the filthy water and tricky currents around Manhattan, looking for sunken ships in the murky mess. Once we found bootleg booze on a tugboat that had sunk during Prohibition. We pulled lobsters from rusty hulks. My proudest moment was when I brought up an old sailing ship's cannon barrel, to the jealous oohs and aahs of my friends. Not until I got it home and scraped off the crust of barnacles did I discover that the artillery piece was actually a length of municipal sewer pipe.

But once the kids were old enough to leave home, I gave free rein to my fantasies. I had accumulated enough money to indulge myself, and I jumped into a whirlwind of diving, climbing, sailing, and racing. My friends and family couldn't quite decide whether I was having a midlife crisis or had just gone completely nuts.

With some buddies from Chamonix, I climbed Mont Blanc, and a very tricky peak called the Index Finger. There were "white spots" here—not the kind on the map, but white glaciers under my feet. On Baffin Island up in the Arctic, I made a survival trek with my son Tony and Fritz. We had an excellent time living in the permafrost off the land and the sea, but here again everything had already been discovered; the white was only previously mapped ice and snow.

I experienced the white spots of foamy water as well. With virtually no sailing experience, I crossed the Atlantic in a 38-foot sailboat with three of my kids. Nina stayed back home, to collect the insurance loot in case we didn't make it. Later I sailed a 47-foot cutter solo from England to the States, in the wrong season, against the prevailing winds. I saw a lot of white water, especially during the frequent gales, but none of it unexplored by man. It was territory new only to me.

One day, back at home in New York, Carlos Casabal, a friend of mine, told me that he was going to race in the famous 15,000-kilometer Paris–Dakar off-road rally. I asked to come along, and he agreed. Eventually he backed out, but like an idiot I decided to go anyway. I hadn't ever raced before, but that wasn't what made me feel the most stupid. What bothered me most was just being a part of this ridiculous spectacle, this money-wasting, self-aggrandizing circus. I was embarrassed to be one of the rich European daredevils shelling out tens of thousands of dollars to go zipping through villages where people were starving, cheered on at every pit stop by emaciated skeletons with distended bellies and eyes full of pus.

We roared through the barren Sahara, leaving behind piles of oil cans and clouds of stinking exhaust. On the nineteenth day of this twenty-one day bout of insanity, my navigator (who drove the car when I needed a rest) missed a curve, and we crashed into a gorge. Since the race organizer himself had died a few days previously (his helicopter had collided with a sand dune), the committee was even more disorganized than usual. They didn't find us for three days. I'd received a nasty gash in my leg, which a local shoemaker had attempted to stitch up with thread, and by the time medical help arrived the injury had turned gangrenous. The French race doctor wanted to amputate my leg, and when I refused to let him, I had to sign a document absolving him of all responsibility if I should die.

While recuperating back home, I mulled over my past few years' adventuring. In retrospect, it all seemed painfully self-indulgent. I had made huge outlays of money for equipment, transportation, and organization, all of it for my own gratification, none for anyone else in the world. I had spent a great deal of time, money, and effort looking for white spaces that no longer existed.

Araouane changed all that. A vast, unexplored, unmapped territory had showed itself to me, a huge white spot of sentiment rather than space. Had I known just how far this unknown terrain would extend into my life, I might never have undertaken my project. I'd had a hard enough time dealing with my own flesh and blood, and I had no intention of adopting a whole new family in the middle of the desert. But each day the people of Araouane began to feel more and more like kin.

The place had gotten into my blood. On my first visit, with Dah, I had stayed only seven days; the second, seven months. The next two years I stayed longer still. I had begun to plan to spend much of my life in Araouane and would probably be there right now if not for certain events which I'll relate in due course.

It was ironic that the townspeople should see the rain as a divine gift brought about by my arrival. One of my toughest tasks in the beginning had been to convince them that I wasn't a missionary. After all, who but a missionary would dole out food, take up residence, and try to teach their children? Experience had taught them to seek an ulterior motive.

During my time in the region I heard tales of the many

Catholic and Protestant missionaries who had come to the Sahara. Almost always, my Muslim friends told me with a chuckle, they left without having made any converts.

"Some pretend to accept Christianity," one friend told me. "They tell the missionaries that they've converted so they can receive free food and education, but they think of it just like a job. After 'work' is over, they go and pray at the mosque."

The patricians who'd left Araouane for Timbuktu were particularly suspicious of me. They kept a quiet watch, waiting to see how long it would be before I dropped my "pretense" of being merely a gardener. Babaya and Araouata monitored my activities in the village and sent reports back to their city cronies. They even had me carry their messages, confident that I wouldn't be able to decipher the arcane form of Arabic they used. I didn't find out I'd been carrying the records of my own surveillance until much later.

Back when I was teaching school, before we hired Mohammed Ali, I sometimes gave the kids lessons in geography. I had an inflatable globe on which I showed them Africa, America, and Europe. I showed Mali's location in relation to neighboring Algeria, Niger, and Mauritania. I pointed out China, and told the children that this was where the green tea they drank came from. One day I tried to explain why they faced east to pray. I showed them the location of Mecca, and said that if they walked a very long time in the right direction they would wind up there. If they walked even farther, I explained, using the globe to illustrate the point, they would have to face west when they prayed. I showed them how if they continued walking in the same direction they would come to America, and eventually wind up right back where they started.

This little fantasy trip seemed harmless enough, but when word of it reached the VIPs in Timbuktu, they did not like it one bit.

"Is it true?" one of them asked me on my next visit to the city. "Do you teach the children of Araouane that they can pray to God by facing west?"

"Did you tell them," asked another, "that the world is round?"

"Do you not know," a third scolded me, "what the Prophet has told us—that when the end of the world comes, the angels will just roll up the stars, the land, the water, and everything else into a big ball and put it away? How could they do that if the earth were round? Are you telling our children that the Prophet was wrong?"

"Do you believe in God?" asked the first merchant.

"Do you worship?" asked the second.

Most of these wealthy men were self-styled marabouts. It added to their prestige.

I began to get regular invitations to dinner in the homes of these patricians when I was in Timbuktu, but it took me a while to catch on that their purpose was to give the host a chance to sound me out on spiritual matters. My children have sometimes accused me of being oblivious to other people's feelings. Perhaps this is true, and lucky, too, because if I'd had any notion of the mistrust surrounding me in those early days I would have become paranoid. As it was, I just sampled the festive dishes and downed countless glasses of tea as I answered the flurry of questions the old patriarchs addressed to me.

There was one question I didn't answer with absolute candor though. When asked whether I believed in God, I'd always reply that I didn't believe in him in any organized way, that I felt whatever God had to tell me He could do directly, without the intermediary of a prophet, a church, or a mosque. I suppose this made them a little more comfortable than if I'd claimed to be a Christian fundamentalist. And it was certainly better then telling them the truth—that I am a firm atheist.

During my second year, however, my carefully constructed cover was blown. My friend Julio, who had come all the way from New York to be the first paying customer in Araouane's hotel, was down in Timbuktu with me on his way back to America. We were having dinner at the house of a merchant named Hadjim, surrounded by other well-to-do Moors. We sat on thick carpets, helping ourselves with the right hand to delicacies from a platter held by a serving boy. Julio noticed a blackboard with Arabic writing and asked our host what it was.

"My marabout is giving me lessons from the Qur'an," said Hadjim. "Are you religious?"

"No, no," Julio answered. "I grew up in South America, in a strict Catholic family, but I gave all that up long ago. I can't believe in the existence of any God who would create as much misery as we have on earth. Either He deliberately makes us unhappy, or He is powerless to prevent evil from coming to us, so I can't believe He is there at all."

I was a bit shocked that Julio would speak so tactlessly in a room full of deeply religious Muslims. But he was leaving town the next day, so I figured it really didn't matter.

"Well, your friend Ernst believes in God," Hadjim said.

Julio burst out laughing. "Is that what he told you? I didn't think he was so stupid. He must have changed his mind out here, since he certainly didn't believe any of that crap back when I knew him in America."

Ten pairs of dark eyes instantly turned to me. I could have strangled him.

"Is that true?" Hadjim asked me. "Is this man telling us the truth?"

"I am afraid he is," I admitted. "I didn't say this earlier because I did not want to offend you, and I do respect your religious belief. But for my part, I do not believe in a personal God. Do not worry, though—I would never try to impose my atheism on other people."

If this exchange had taken place a year earlier, I might as well have packed my bags on the spot. Fortunately, by this time the men of Araouane and Timbuktu knew me well. I had earned their trust, and they were willing to overlook my godlessness. They talked among themselves for a few minutes, and then Hadjim turned to me with a wry smile.

"Since we know that you really respect our belief," he said, "we respect yours as well."

"But," another man added, "we may try to change your mind."

Over more cups of tea, the Moors told me all about their fears that I'd turn out to be a missionary. They'd been especially concerned about reports that I was making the schoolchildren sing Christian hymns. I had never done anything of the sort, so I asked them which hymn they thought I'd been teaching.

"We do not know the name," said one of the Moors, "but we are told the tune goes like this."

He hummed a few bars, and I immediately recognized it. The so-called hymn was indeed a song that I'd used to teach the pupils in their French lessons, and I even knew the words: "*Alouett-e, gentille Alouett-e. Alouett-e, je te plumerai . . .*"

Araouane did not make a missionary out of me, but it did turn me into a doctor. With absolutely no medical training, I soon found myself dispensing advice and treatment and even performing minor surgery.

One day during my first trip with the salt caravan, a nomad had appeared with a little boy on his shoulder. It was obvious why he had sought us out: one of the kid's knees had become a swollen red bulb, and the local people assume that any *toubab* is necessarily a doctor. Had the boy been one of my own children I would have tried to puncture the cyst and drain the pus, but without any way of telling the nomad how to prevent infection I didn't dare open up the knee. Instead, I asked for some boiling water and salt, and made a hot saline compress out of one of my T-shirts. I gave the father some of my scant supply of antibiotics, cutting each pill in half to stretch out the treatment.

Word spread like lightning through the area, and a stream of sick people began showing up at our campsite every night. The most common complaint was a thorn lodged in the sole of a person's foot, so my pocket knife, tweezers, and little bottle of alcohol were constantly in action. Dah had the customary obligation of serving tea to every visitor, and sometimes he poured without a break for hours on end.

During all my time in Araouane, my medical duties never let up. At any time of the day—as I was cooking my dinner, while I was teaching school, even after I'd gone to bed for the night—people would show up at my doorstep with ailments that needed tending. They were fascinated by "quinine" (as they called all Western medicines), and would often concoct wholly imaginary illnesses just to get some pills. Whenever I gave a nomad some medicine, his comrades would get jealous and decide they needed the same treatment. They'd all cough, wheeze, grunt, and moan until I got rid of them with a brightly colored vitamin pill.

The villagers were great hypochondriacs as well. Boudj's mother, for example, seemed to have some new complaint every time I saw her. She was just the opposite of her hard-working son: lazy, bad-tempered, and constantly whining about some problem or other. When she interrupted my dinner to demand aspirin one too many times, I decided to teach her a lesson. I said that since she was in so much pain so often, I'd give her a medicine more potent than mere aspirin. I gave her a tablet of Chloroquine, a harmless malaria medication. Be sure not to swallow the pill, I told her, but rather suck on it slowly until it dissolves. She popped it into her mouth and made a grimace; Chloroquine is one of the most vile-tasting substances in the world. She never complained of any ailment again.

One of the nomads who sometimes joined Dah's group when I was with the caravan was known as Crazy Sidi. He had been born a number of years before Dah, "in the year when the locusts ate everything up and all the goats died," and had lived all his life in the same region of the desert. Yet he still could not find his way from one watering hole to the next without following somebody.

Crazy Sidi was amazed by all the modern gadgets I'd brought with me. The flash on my camera mesmerized him for hours. I showed him how to work the test-button, and he would sit there flashing and recharging the device all night. He loved the whir of the automatic shutter advance. He'd put the camera up to his ear, or peer into the lens trying to see what sort of animal trapped inside was making this strange sound.

Crazy Sidi's wife, however, was more interested in my medicine than my technology. She was no more than half his age, and very outgoing. She confided to me that she would like to have some more children. If I could create light from darkness with a magical flash box, who knew what other miracles I might have up my sleeve?

When I asked Crazy Sidi if he would like more kids, he made it abundantly clear—through rather graphic sign language, since my vocabulary certainly didn't include the requisite anatomy or biological processes—that he was all for the idea, but there were some technical problems. I gave Sidi and his wife a couple of my colorful emergency vitamin tablets, an aspirin, and some glucose pills for instant energy. The nomads are convinced that "quinine" never fails to work magic, and I am convinced of the power of positive thinking.

The next morning Sidi came up to me with a big grin on his face.

"They worked!" he said. "Last night we made a baby!"

Out in the desert, the traditional remedy for almost any ailment is dried camel shit. A runny nose is cured by shoving an appropriately shaped turd into the dripping nostril. An upset stomach is soothed by an elixir of dung mixed with water. An open wound is salved by a paste made from shit and ashes. The wounds actually heal, some of the time, but I don't know whether it's because of the potions or in spite of them.

Occasionally I'd see a traditional remedy that seemed to make sense. One herder, for instance, had broken his collar bone, and his whole upper torso was caked with a crust of dung and camel hair hardened with resin. This bristly plaster looked rather gruesome, but at least it immobilized the injured part until the bone could fuse. I couldn't imagine how the cast would ever be removed—maybe the locals make a solvent out of camel manure as well.

One evening, Emilie and I were sitting outside my house at sunset smoking. It was a habit we shared with all the local men except Mohammed Ali, who had been running a campaign to keep the kids from taking up that curse. Once, while I was in America, the local men had run out of tobacco, and they'd dispatched Habbabou with a caravan to Timbuktu for emergency supplies. During the two weeks he was away, the other villagers nearly went crazy. They took to smoking tea leaves, chili peppers, chicken feathers, anything and everything they could light up. They always had somebody posted on the highest dune to keep a lookout for the tobacco relief caravan.

"Do you want to become like that?" Mohammed Ali asked the children. He made his point, and not only didn't they take up the habit, but they'd taken to chiding us for it.

A young woman named Agida practically crawled up to my house as we sat there smoking. She was usually one of the most cheerful women in the village, but she clearly had an earache again. It was a chronic ailment of hers, but this time it seemed worse than ever; the whole side of her face was horribly swollen.

I took a look, and found her ear stuffed with some sort of solid matter. I asked what it was.

"It's sugar, tea leaves, resin, tobacco, and ground camel turds," Boudj translated for her.

I decided to try and dig the stuff out, so I could wash the ear with chamomile and alcohol. I tried with cotton swabs, then tweezers, but the mass would not budge and the ear started to bleed profusely.

"You'll have to soften it up first," I said through Boudj. "Soak the ear in warm water, then wrap it in a hot cloth for the night. Once it loosens up, we'll clean it out and apply medication."

Agida thanked me and got up to go.

"One more thing," I said. "Let me give you some antibiotics—the right side of your face is so swollen there must be a serious infection."

I started to get the pills, but Boudj called me back.

"Wait, Ernst," he said. "I think you are wrong about that infection. That is not swelling, it's camel dung."

I asked Agida to open her mouth. As Boudj had suspected, the right side was full of shit.

"What on earth does she do that for?" I asked. I thought I'd discovered all medicinal uses of camel dung, but this was a new one.

"Usually it is placed in the mouth only for toothache," Boudj explained, "but this time the pain in her ear was so strong that she put the remedy everywhere she could."

Before my arrival in Araouane, the only medical treatment (if you could call it that) was a combination of faith healing, genuine folk remedy, and God-knows-what-else practiced by the holy men. They based their fees on their reputations and the effectiveness of their cures—a good marabout was literally worth his salt.

Each marabout had his own secret methods of treatment. Araouata was strictly a Scriptures man—he would read passages from the Qur'an until the patient fell asleep. Sidi (not Crazy Sidi, the nomad, but the local man of the same name) sold gris-gris, little leather pouches with Qur'anic verses sewn inside, to be worn around the neck. Old Baba Cambouse specialized in wrapping little pieces of string around the diseased body part, but to treat his own ailments he relied almost exclusively on camel dung.

The proud old man had a permanently dripping nose, and complained of constant body aches and general fatigue. I

suspected that it was simply the kind of affliction that normally comes with old age, so I gave him some iron pills and
painkillers from my pharmaceutical supply. I had also noticed that he had to squint whenever he tried to look at
something close-up, and suspected he was terribly far-
sighted. I'd brought two pairs of reading glasses to Araouane
for my own use, and gave one to him. For a few days he
wore them all the time, even when trying to see things off in
the distance, but then he abruptly stopped. I asked him why
but couldn't get a straight answer. I concluded that he must
have sat on them accidentally and been too embarrassed to
say so.

Generally, the villagers seemed to be in excellent health.
The little children sometimes had coughs, but no more often
than American kids. The most common complaint was
toothache, and all I could do for that was hand out aspirins.

One day a woman named Hadja came to me to get medication for her baby. We didn't see her often at the garden,
since her father was a rich Timbuktu merchant who always
kept her supplied with food. It was rumored that Hadja had
recently been abandoned by her husband, but nobody
would speak openly about such things to avoid embarrassing
her. Her baby wore a sort of crown made of fabric, but the
headdress circled the face vertically rather than horizontally,
as if to hold the child's jaw in place. At evenly spaced intervals, small oval lumps were stitched into the cloth.

"I hope you have some medicine for the little one," Hadja
said.

"What's the matter with her?" I asked. "She looks fine."

"No," Hadja replied, "she cannot hold her head up properly."

"But she's holding it up now," Emily said.

"That's only because of the band that the marabout made
for her."

We did not ask which marabout had given the headdress.
All the holy men in town guarded their secrets closely, and
each was deeply jealous of the others.

Emilie gave Hadja some vitamin pills for the baby, and took a closer look at the headband. After examining it she found that the evenly spaced lumps sewn into the fabric were—of course—camel turds.

No real rivalry developed between traditional cures and modern Western medicine. The villagers wanted to hedge their bets so they'd go to me for miracle pills or elixirs and also to a marabout for spiritual healing. The marabouts themselves asked me for "quinine" whenever they had a pain. And perhaps, in the final analysis, my own cures were more spiritual than medicinal: Most of the pills I gave out were just vitamins or aspirin, and the biggest, most colorful ones always seemed the most efficacious.

Sometimes during my term as doctor I had to be my own patient. During my second year, just a few days before I was due to make my trip back to the States, I woke up one morning with an intense burning sensation just above my right ankle. I looked down and saw a pale 3-inch-long scorpion crawling over my sleeping bag. I crushed it with a flashlight.

I knew the old theories on poisonous bites and stings, and the new ones as well: Current thinking says you don't cut the affected area to let it bleed, apply suction cups, put on a tourniquet, or wash the limb in vinegar or alcohol. Instead, you simply wrap the site as tightly as you can, refrain from coffee and liquor, and stay immobile until you can get to a doctor.

The only problem with this approach was that I *was* the doctor. I sat and considered the situation.

First, my Land-Rover was the only car in town, and at this point I was the only person who knew how to drive it. Getting to a real doctor would be no easy task. Second, if by some miracle I managed to drive myself to Timbuktu, I could hardly expect good treatment at the hospital; from all accounts, it was a place where people reportedly stole the ears off your head. Most medicine is sold on the black market before it reaches the stockroom, so the doctors there can offer you almost nothing but advice. Third, I had told the villagers many times that we could take care of all our own problems ourselves, without any outside help.

I've always believed that a positive state of mind can cure most physical ailments, so I looked around for something to improve my mental state. The only thing on hand was coffee. After putting on a rough bandage I went to the kitchen and brewed a pot, even though coffee itself and the effort necessary to make it were both on the list of poison no-no's.

My leg soon became incredibly heavy, and felt boiling hot all over. A huge burgundy-colored rash developed, as if I'd spilled a bucket of paint on the limb. The site of the sting itself did not hurt so much anymore—once the pain was evenly distributed it became much easier to bear.

I laid both legs on the table and sipped my coffee. I remembered a casual conversation I'd once had with a Belgian toxicologist who had lived for many years in the Sahara; there were regional differences, he'd told me, between scorpions of the very same species. In the region of Tessalit, 400–500 kilometers from Araouane, the white scorpion's sting was almost always fatal. In the region of Taoudenni and Araouane, however, the poison was far less malign. Almost no problem at all, my mind told my body. Almost no problem at all! Almost no problem at all! Straight from the mouth of a scientific researcher! I repeated it over and over, until I started to feel better. I felt particularly glad to be in Araouane and not Tessalit.

For the next few days I had no fever, not much pain, no big problems other than a horribly swollen leg. I tried to walk around as if nothing had happened. The kids were proud of me when I told them what had happened. "We make a much bigger fuss when somebody gets stung," they said.

Boudj spread the word that I, who gave out medicine all the time, didn't take any myself. He vowed to follow my example and never take medicine if he got sick. "Ernst can cure everything with his mind," the boy said, "and I am going to do the same."

I don't know if refraining from taking medicine was the right course of action, but I've long believed in the healing powers of the mind. I don't have contempt for "real" doctors, but I feel we are too quick to run to them whenever we get sick. I tried to instill this conviction in the children of Araouane, so that they wouldn't become psychologically dependent on Western medicines that might be unavailable after I left. Mohammed Ali felt the same way, and shunned both pills and gris-gris, except for an occasional aspirin for toothache.

If I had been staying a bit longer in Araouane, I could have kept my resolve, but the infection got really bad. When I walked around, it felt as if my leg was going to fall off, and it swelled up so big that I could barely get my pants on. I thought about taking an antibiotic without telling anybody, but that would have been cheating—a thing we didn't do in Araouane. And soon I would have to drive to Timbuktu to make my flight to the States. In the end, it was Boudj who persuaded me to give up my demonstration of the mind's healing powers.

"Ernst," he said, "you cannot drive with your leg so swollen. How about taking some medicine? This is a special case, since you have to travel. I know you can heal things with your mind, but antibiotics will make it go faster."

I took the medicine, which eased the swelling enough that I could make it to Timbuktu. But I did not take my chances at the local hospital, and did not see a "real" doctor until I reached the States. By then the swelling and infection had gone away, so she couldn't evaluate my improvised treatment.

"Whatever you did seems to have worked," she said.

I could have figured that without paying for a consultation.

I not only became Araouane's doctor, but the town's resident authority on all matters of modern technology. People assumed that because I was from the West, I knew everything about Western machines and equipment. Not only the villagers, but all the nomads of the surrounding area regarded me as their mechanic, engineer, radio repairman, and even gunsmith.

One evening, while Emilie and I were preparing dinner, four nomads came to the door. One was carrying an aging Kalashnikov rifle.

"Shit!" I muttered. "What are we in for now?"

But they were not here to steal or cause trouble. "Aebi," said the leader, "we have heard that you can repair anything. This gun does not work anymore. Could you please fix it?"

"I don't know much about guns," I said, "but let me have a look."

A bullet had gotten stuck in the barrel, and when the nomads had tried to dislodge it with an iron rod they'd ended up getting the rod jammed in there as well. As I sat outside my house fiddling with the weapon, a crowd of nomads and villagers gathered to watch. Many of them had already tried to fix the gun, and it seemed like an impossible task to them.

With slow, patient tapping I managed to get both the broken rod and the bullet out of the barrel. The nomads' admiration was boundless. The Kalashnikov's owner promised me the first gazelle he shot with it, which I declined. I didn't want to encourage the killing of these graceful animals—and the boost to my reputation as a jack-of-all-trades was thanks enough.

I suppose I'd always had a knack for tinkering, long before I began studying engineering. More importantly, my parents had taught my brothers and me that we could accomplish anything we set our minds to if only we tried hard enough. Therefore, throughout my life I've often taken on tasks that were far beyond my abilities. Sometimes I've completely embarrassed myself, but usually I've managed to muddle my way to a solution.

When I was thirteen and my brothers were eleven and eight, my parents bought an old abandoned house in the mountains. It was unreachable by car, since a large cliff separated it from the main road. My father gave my brothers and me a box full of plastic explosives, caps, and fuses, and told us to blow up the cliff.

"How do we do that?" I asked.

"I think you stuff the plastic into cracks in the rock," he said, "then stick the caps into it, attach the fuses, light them, and run away." He shrugged, and went off to work in the house.

The cliff was destroyed, we all survived, and on Monday all our schoolfriends were green with envy.

So I may not have had much experience fixing engines or tampering with jammed rifle barrels, but by the time I got to Araouane, I had spent over forty years tinkering with problems beyond my abilities and teaching myself along the way.

Araouane not only gave me new professions, it gave me a new name.

From my first visit to Timbuktu, the city's throng of little and not-so-little street kids hounded my every step. They were as noisy and persistent as a swarm of mosquitoes, but I grew just a tiny bit fond of them after a while.

"Ernst," one would say, tugging at my sleeve, "I'll show you the restaurant."

"Ernst," another would chime in, "give me a present."

"You need oranges, Ernst? No problem, come with me, I know where to get them."

"Please, Ernst, a thousand francs for my poor mother."

"Want a woman, Ernst? I can set it up."

"Ernst, trade me your shirt for this dagger. See, it's antique."

"Ernst, give me a thousand francs."

It was the same for any Western tourist who showed up in Timbuktu. The urchins had a talent for remembering names and faces.

When I returned the second year with my trailer full of saplings, I found I had been promoted from the rank of tourist to honorary resident. The little mafia didn't pester me quite so much, but whenever I asked one of them to run an errand for me, a fistfight still broke out over which young hoodlum would be the one to earn the commission. They now called me *"Cher Excellence."* (Where they got that expression from was beyond me.) Sometimes they even referred to me as *"L'homme d'Araouane."*

In the third year I drove from Reggane to Araouane direct, without a guide other than Emilie, to avoid areas then brewing with the Tuareg rebellion. It was a fairly dangerous trip, one that seasoned Sahara hands were reluctant to undertake. When the little gangsters of Timbuktu heard about it, I rose far higher in their esteem. Not only did they give up dogging me for handouts, but when I needed a service from them, they accepted my offered price without subjecting me to the usual drawn-out session of haggling. Their newfound courtesy extended to any friends or relatives of mine who happened to be passing through town.

What's more, the street kids now had a new name for me. I was *"L'homme du desert"*—the Man of the Desert. I liked the sound of it.

During my last year in Araouane, I decided to prove myself worthy of my new name. I was amazed that the villagers spent their whole lives surrounded by the desert, yet most never ventured out in it unless they had to. They saw the desert as a place to escape from, not somewhere to seek.

Once a woman of the village went out looking for dung in the morning, and by evening she still hadn't returned. Although she'd been gone only a matter of hours, the other townspeople considered her as good as dead: How, they seemed to feel, could anyone survive in such a wasteland? We sent out a search party, on foot and camel, and Fritz with the Land-Rover, but could not find any trace. Without powerful flashlights or the help of a full moon we had little chance of finding her that night.

In the morning we went out again, but the villagers seemed to feel it was barely worth the trouble. One might

have assumed that they could follow tracks in the sand; there
had been practically no wind to obliterate them. But nobody
in town knew how to read tracks. Nor did they seem to
know anything about desert survival. The nomads saw the
barren dunes as their home and could read the terrain like a
map, but most of the sedentary population of Araouane was
no more familiar with the desert than I was.

By noon most of them had given up the search as a waste
of time. She had no water, they said, so why bother looking
for a dead woman? I practically had to force Moulay Mokhtar
to go out with Fritz and keep searching, but it was to no
avail.The woman's mummified corpse was found two months
later, by passing nomads.

I firmly believed that a person could survive in the desert
for quite some time. It had rained a few months ago, and big
patches of *halfa* grew in many areas. This desert grass, I had
noticed during my time with Dah's caravan, had juicy deposits
in its stem that could, I was convinced, sustain you in an
emergency. So I made a challenge to the men of the village:

"I'm sure it's possible to live in the desert," I said, "without
bringing any food or water."

They all began to laugh.

"You must be joking," one said.

"I shall prove it to you. I'll just go out for a couple of days,
take nothing with me, and if I come back okay, I'll have made
my point."

"You cannot do that, we won't let you go," old Baba Cambouse said.

"Don't worry," I said. "Fritz will drop me off at some remote site, and every day he will come to check on me. Surely
I won't turn into a corpse overnight."

The villagers did not hide the fact that they thought I was
crazy. Hamd'r Rahman, the nomad who often hung around
town, was incredulous. He could do it, he said, but he very
much doubted that I could.

Fritz drove me out in the Land-Rover. I selected an area

about ten kilometers from the village, where quite a lot of *halfa* grew. I'd brought a tent, a sleeping bag, my short wave radio, and (of course) my Swiss army knife. The idea was not to rough it, not to live without any comfort at all, but simply to see if I could survive without carrying food or water. If I found I was in desperate need, the village was only two hours' walk.

After setting up the tent I went to pasture. Each shaft of *halfa* had a juicy tip no bigger than the fingernail of my pinky, so gleaning nourishment from the grass was hard work. It took a good bit of experimentation to find which type of stem had the most juice and the edible pulp. On the first day I had to tear apart *halfa* stalks for over seven hours to consume enough pulp to sustain me comfortably.

The second day, however, it took me only four hours to glean the same amount of pulp. By the third day I'd become expert enough to cut the grazing time down to two hours.

After three days and three nights in the desert without a single gulp of water, I wasn't in any distress. I was urinating copiously, which meant that I was in no danger of dehydration. I wasn't a bit hungry, since the grass pulp provided my food as well as drink. The only problem was that I had gotten bored out of my wits.

I had assumed that I would have to fight to sustain myself, only to discover that the whole thing was child's play. I had come simply to survive, and found mere survival too simple to occupy my time. I got tired of watching ants, spiders, *gang-gang* bugs, lizards, and birds—any of which I also could have eaten, in a pinch.

When Fritz came by on the third day, the Land-Rover packed with curious kids, I had had enough. I had easily survived three days in the desert, and now I had better things to do. But before I hopped on board, I gave the most irrefutable proof of my success I could think of: a long trail of piss in the sand.

VISITORS

I originally conceived of the hotel merely as a way of providing some protection for the windward side of the garden. Araouane's population had been dwindling for decades, and the abundance of abandoned houses made building new ones seem like an exercise in futility. But none of these derelict huts would have been suitable to lodge tourists. Through most of its history, Araouane had prospered by providing services to travelers in the Sahara. Today, with salt caravans growing fewer and poorer all the time, the only travelers with money to spare were the European tourists and adventure seekers in Land-Rovers. No matter how well the villagers learned gardening, they would always need at least a little cash income to buy necessities from the outside world. I came to realize that a hotel would be vital to the town's success.

I didn't particularly like the idea, since tourists generally have an unhealthy influence on traditional societies. Vacationers just "letting it all hang out" often assume that their hosts will cater to their every whim, and they show little concern for the local population. But any travelers who made it to this forlorn place would probably be a special sort—resourceful, self-reliant folks who didn't mind working harder during their holidays than during their office hours. Their determination and sense of purpose might even be a good influence on the villagers.

Around Christmas time of my first year, construction began on the Araouane Hilton. I had no feasibility study, no committee proposals, no comptrollers, no investment analysts, and since we hadn't yet a common language, I didn't even discuss the plan with the villagers. I selected the hotel's name in hopes that the Hilton chain would get wind of it and give us some free publicity by suing us.

Of course I talked it over extensively with Mohammed Ali, who had already gained my full confidence. We decided to build ten small rooms, exclusively with local materials. The shelves and tables would all be made from salt bars. The hooks for hanging clothes were to be goat and gazelle horns, the floors a mosaic of broken pottery which the children could gather in the sand around the village and the beds raised sand boxes with mats of woven *halfa.* We would have to make concessions with the roof beams, doors, and windows and bring these up from Timbuktu, because of the lack of wood in Araouane. Since even Mohammed Ali was unable to decipher my plans, I drew the outlines of the foundation in the sand.

Together with the workers from Timbuktu, the children, and the old villagers, we muddled our way through the construction. Invariably when I turned my attention to some other activity, I would return to our Tower of Babel only to have a new wall torn down because a future room had no door, or because its dimensions would only allow future occupants to sleep upright. The maze of emerging cubicles, courtyards, bathrooms, dining room, kitchen, store rooms, forecourt, and patio baffled the whole construction crew.

On my next return from the States I brought Fritz, my sculptor friend with the wild imagination, to help with the final touches and decorations.

The hotel's first guest (and almost all who followed) proved that I wasn't the only outsider to find Araouane's inhabitants delightful. Julio was a native of Montevideo whom I had met at a party in New York the previous summer. He owned a very successful restaurant in SoHo, and in the course of idle chatter I mentioned that I'd be spending most of the next year in Araouane.

"Where is that?" Julio asked.

"About two hundred and fifty kilometers north of Timbuktu," I answered.

"Great," he said, "I'll come."

He did. One day, when I was in Timbuktu haggling with an Arab merchant named Abdi over the few tons of rice he'd stolen, Julio showed up at the door. He had simply caught a plane for Timbuktu and asked the first kid he'd met where to find Aebi.

"Hi, Ernst," he said casually. "When are we going to Araouane?"

His timing was very lucky, since if I hadn't happened to be in Timbuktu, he'd have had a very difficult time reaching the village.

"We go up tomorrow," I said. "You are very welcome to come with us, but once we leave, there will be no way to get back for about a month."

"That's no problem. My partner is taking care of the restaurant."

"You will have to pay for the hotel room."

"Of course, that's fine with me."

"The hotel isn't quite finished."

"No problem."

"What are you going to do in Araouane for a month?"

"I'll figure out something."

I wasn't sure how the villagers would react, since they'd never really understood the concept of the hotel; visitors in the desert are always offered lodging as a matter of hospitality, and the idea that somebody might willingly pay for a service that had always been free seemed ridiculous to them.

When we roared into town, our truck loaded down with rice and baobab powder, the villagers formed their usual welcoming party. Women ululated, kids shrieked, men smiled, hands stretched out in greeting. Then they caught sight of Julio.

"Hi everybody," I said in French. "Greet Julio, our first guest."

The men stepped back and dropped their outstretched arms, and the women simply vanished. Only a few of the children, who'd endured my endless stories about strange lands and strange peoples, ventured uncertain smiles and greetings. With typical South American ebullience, Julio jumped off the truck and tried to gather them in a bear hug. That was a bit too radical a departure from the traditional litany of greetings.

But the children overcame their initial hesitancy, and I don't think it took a day for them to make our visitor feel like part of the community. He and they spent hours on end running around in the dunes like a bunch of wild animals, playing catch or hide-and-seek. Soon all day long I'd hear, "*Julio, viens ici,*" "*Julio, tu veux . . . ,*" "*Julio, montre moi . . .*" This kind of play helped the children learn French at a speed that seemed almost impossible. Julio made watercolor paintings of the village and the garden, and taught some of the kids to sketch.

He did stay a month, and he loved every minute of it. Only one hotel room was finished at the time, and construction was going on all around him. Fritz, may he be blessed forever, was putting his aesthetic sense to use by directing much of the work himself, and the children painted wild frescoes on all the walls with different colors of banco. Since there was not much else to do, Julio often lent a hand on the construction.

I wanted to show the villagers what the hotel could mean to them, so at the end of Julio's stay I had him pay his bill in front of everybody. The money itself (300,000 francs, or about $1,200) still didn't mean much to them. But when I told them this sum would buy 3 metric tons of millet, so much that it would take thirty camels to transport it, the townspeople were finally sold on the Araouane Hilton.

Not all our visitors stayed in the hotel. One day a bunch of kids came running up to the garden, scared out of their wits.

"Help!" one of them screamed, "everybody, come quick!

There is a crazy man in the village who wants to kill all the goats! He also says he'll kill any person who comes near him."

We dropped our gardening tools and ran back to town. I scrambled to my house for a chain in case we needed to restrain the intruder. Some of the men were already converging on the village center, with shovels and hatchets in their hands.

By the time I arrived a big crowd had formed in front of Habbabou's house, and in the middle stood our madman. He was a diminutive fellow in a ragged French army greatcoat that reached almost to his feet. His tanned face had vaguely Mediterranean features, framed in a tangle of bushy, snow-white hair, and bore the expression of a cornered animal.

I asked him if he spoke French.

"*Bien sûr,*" he replied. "I am the King of the Burned Earth, the Prince of the Promised Land."

"What are you doing here?"

"I am joining my armies, which are some distance to the north."

"Where? In the salt mines?"

"Of course not," he replied contemptuously. "My armies are awaiting me there." He pointed to the northeast.

By this time I had determined that our visitor wasn't any danger to the town, and I told the villagers that they could go back to work. Most of them were bound by curiosity, however, and stayed to watch me talk to the visitor. This was the most excitement they'd had in quite a long time.

"When you have your army," I asked, "what will you do?"

"I shall reclaim my rightful kingdom of the Burned Earth."

"From whom?"

"The country next to America."

"What? From Canada or from Mexico?"

"No, no," he said, with growing impatience, "from the country *between* Canada and America."

"I see," I replied calmly, "but in that case you are walking in the wrong direction. You must travel south until you reach

Timbuktu, then turn right and walk west until you reach Dakar, and then find a ship, because there is a huge ocean between Africa and America." If he continued north from Araouane as he intended, he would surely die in the desert.

"Hmm," he pondered for a minute, "perhaps I should inform my allies."

"Who are your allies?"

He reached into his coat pocket and pulled out a big wad of folded papers wrapped in a piece of burlap.

"Look for yourself," he said, "they all know I have been cheated out of my rightful inheritance. They will all come to my assistance."

He handed me the papers. They were letters, clumsily scrawled in French, addressed to the kings of Egypt, Sudan, Syria, Pakistan, Indonesia, Iran, Iraq, Malaysia, and Afghanistan. The King of the Burned Earth had sought aid from the imaginary monarchs of every Muslim nation he could think of.

"As soon as my allies arrive," he said, "I'll retrieve my treasure, which is buried in the desert to the north."

He told me he'd come from Adrar, "mostly on foot, but sometimes with detachments of my army." I couldn't believe he'd made it on foot from a city 1,500 kilometers away, through a part of the desert that contained no wells and no nomads, without so much as a canteen of water. But if he'd arrived from the south, he surely would have run into some nomad or other, and we'd have heard of it almost immediately through the mysterious Saharan gossip mill that somehow carried news across miles of empty desert.

From his appearance and accent, I guessed he was Algerian, but I couldn't get a straight answer. We tried to persuade him to stay and rest a bit, but he was determined to go join his phantom army. We gave him a sack full of grain, dried camel meat, tea, and sugar, as well as a plastic jerrican of water. We showed him the direction to Timbuktu, and he calmly walked off that way.

Two days later, while we were all working in the garden, someone noticed a figure moving in the desert.

"The King of the Burned Earth!" our lookout cried, and pointed east.

Far in the distance, we saw him plowing through the dunes in his tattered greatcoat. He had doubled back and was heading north now. We shook our heads, knowing he would not find any water or food in that direction for almost 2,000 kilometers. Unless, of course, he really did have an army waiting for him after all.

About a month later, cameleers returning from the salt mines reported that a white man had shown up at Taoudenni. He had come out of the desert, asked for a bit of water, refused offers of food, and then vanished into the wastes again.

Maybe he really was the King of the Burned Earth.

Another time, a caravan coming from Mauritania brought us a present. It was a very tall man, buck naked, with pitch-black skin. He had been found wandering in the no-man's-land between Araouane and the Mauritanian town of Oualâta 400 kilometers to the southwest.

"He cannot be a bad man," they said, "because he prays five times a day. We have tried to speak with him, but he doesn't understand any language we know."

I tried French and English on him, since many people in this part of Africa know at least a word or two of some European tongue, but the big man gave no sign of comprehension.

First we gave him some clothes, because the prudish Araouanites were uncomfortable with this towering man who wore nothing but a vacant grin. Sidi Mohammed's few words of Hausa elicited a sentence of response, but the guide didn't know enough of the language to actually converse with our visitor.

The villagers, as was their custom in a new situation, stood around the big man in an uneasy circle, torn between curiosity and fear of the unknown. They chattered nervously as they examined the apparition from a safe distance. We offered him some water, which he drank greedily. Mohammed Ali and Bou-djema brought him some supplies from our warehouse, and the man ate the rice and millet raw.

We used sign language to welcome him to stay in the village for some cooked food, and he seemed to understand immediately. But instead of sitting still while the meal was prepared, he started going from house to house begging. He turned the pair of pants we'd given him into a vessel for carrying grain. He tied one pant leg around his thigh with a piece of string, and poured all the millet he could collect into the makeshift sack.

He stayed several weeks, living in a little camp he'd set up in a dune near the garden. Every day he'd beg for food, hurry back to his campsite to scrape it out of his pants, and eat it, then walk back to the village to beg for more. We tried to get him to carry some banco bricks or to help in the garden, but there was no way we could entice him to do any work at all. He was dirty and smelled atrocious, and the only activities he seemed capable of were begging and eating.

The children found it all very amusing, but I couldn't wait to get our guest back to Timbuktu. I did not relish the prospect of sitting next to him in the Land-Rover all the way to the city. Besides, I worried that he might not want to leave. Here in Araouane he could get food all day long without having to do anything for it, so why should he consent to get in the car at all?

The next time I had to go down to Timbuktu, I crept up on our guest with a rope in my hand. I expected I'd have to tie him up for the journey, but he was remarkably cooperative. He got into the car as peacefully as a lamb, and since I had a few miners riding down with me, I didn't have to worry about the tall man causing havoc if left unfettered in the truck. We let him out in Timbuktu, and the giant vanished into the crowd without a word.

Later that year, while shooting the breeze with some friends in Timbuktu, I heard that a flier was being circulated about a tall Hausa man. I asked for the details.

"Oh, he's some demented Nigerian," a friend told me. "A delegation has come here searching for him. Apparently he is a very wealthy man who went crazy and is now wandering about like an ordinary lunatic. His family is offering a reward of twenty thousand U.S. dollars to whoever brings him in."

Only two words came to my mind: "Oh, shit!"

By the time the hotel was completed, my rapport with the villagers had grown so strong that we felt almost like family. The presence of my old friend Fritz proved a big help. When I lost my patience or began to grow irritated, Fritz was there to inject a little humor. Once he found a red-and-white-striped union suit in the pile of hand-me-down clothes we'd brought from Switzerland. He told a teenager named Girage that it was the height of European fashion, and the boy eagerly put it on. Girage strutted proudly through the town, the skimpy undergarment clinging to his body like a second skin.

The union suit fitted his body a bit *too* snugly. When some of the women told him how good he looked in his new outfit, Girage started to display an erection that would have made any stud donkey proud. The people of Araouane were accustomed to long, flowing robes, so the boy didn't realize his arousal was clearly visible to all. When the other villagers noticed and began to laugh, Girage was so embarrassed that he forgot to put his tongue back in his mouth. Fritz teased him mercilessly, chasing him around town and yelling that he'd tie the poor fellow's tongue to his pecker.

One evening, in March of my second year, I was having dinner with my friend the governor. I mentioned that I still hadn't been able to visit Taoudenni, and he said he'd be happy to help. He cut through all the red tape and got me an official permit. On the afternoon it was granted, I was so happy I wanted to shout out my good news to the whole world, but the only person on hand was an old man in the lobby of the hotel.

"I just got authorization to go to the salt mines!" I told him.

"That's wonderful," he said. "Will you take me with you?" He had a very heavy Italian accent, and a nervous tic that made him wiggle his nostrils like a rabbit being teased with a carrot.

"That might be a little difficult," I replied. "I live in Araouane, and will be leaving directly from there. Tomorrow I'll drive my truck back to the village, and I don't know how many weeks it will be before I take the Land-Rover up to Taoudenni."

"All that sounds fine. I don't have to be back in Italy for eight months."

"The hotel in Araouane is rather expensive, and once there you wouldn't be able to change your mind and come back to Timbuktu for at least a month."

"No problem," he said, sniffing heavily.

"The price of transportation to Araouane is one goat—"

"Just tell me where to buy one."

"But you have no permit to go to the mines."

"Surely your work authorization would allow for an assistant, so I could be your assistant."

"It's a very uncomfortable trip—"

"No problem, I'm used to hard travel. By the way, my name is Paolo."

There was obviously no way to get rid of the old man. By that time he was sniffling so vigorously that I was afraid he might pass out. His nose fluttered frantically, and his Adam's apple danced up and down his long turkey neck.

He used to be a banker in Milan, Paolo told me, but was now retired. He still managed a ski rental store in the Italian Alps, but a lack of snow had prompted him to shut up shop and come to Africa on a lark.

"What time will we leave?" he asked and started talking in a steady streak. "I'll be ready. I have only a small knapsack. I don't eat much, so I won't be a problem for food. Can I offer you a beer? I've been in Africa for two months now, mostly in Tunisia and Algeria. Usually I hitch rides on trucks, but I came to Timbuktu on a pirogue up the Niger from Gao. The boatman was so stupid, he got the boat stuck on every sandbank, so I just walked along the shore until I could catch another ride. Don't worry, I have money to pay for things. By the way, what is your name?"

At least it seemed that Paolo wouldn't be a demanding guest—if I could just figure out a way to keep him quiet.

When we got to Araouane, I introduced him to the townspeople who worked in the hotel.

"This is Paolo," I said. "He is going to stay in the hotel for a long time. He will teach you how to treat tourist guests properly, so you follow his instructions. I've told him to complain about even the smallest thing that you do wrong, so that you can learn."

The five kids who made up the hotel staff took turns waiting on Paolo. There were no other guests at the time, and there hadn't been many up until this point, so the kids were still thrilled with the novelty of it all, even though Paolo worked them like a drill sergeant.

"What did you wash this glass with?" he yelled at one kid.

"Sand."

"Are you crazy? Sand?! This is truly ridiculous! Glasses are washed with water."

The boy came back a few minutes later.

"Paolo, is this okay now?"

"What did you wash it with this time?"

"Water, just as you said."

"No soap?"

"You didn't mention soap."

Etcetera, etcetera.

Paolo was so funny and loony you couldn't help but like him. His belongings consisted of two pairs of pants, two shirts, and two pairs of underwear, a little sack of toiletries, a camera, and a sleeping bag. Bou-djema told me that every time he gave the hotel kids his pants to be washed, he first unstitched a pouch full of money and sewed it into the waistband of his clean pair. Even though he spoke French fluently, he insisted on teaching the children a few words of Italian. Soon the kids were chirping out *"pasta," "buon giorno," "ciao,"* and, as their teacher often exclaimed, *"che bestia!"* which loosely translated means "what a jerk!"

Maurice and Viviane, my friends from the Belgian relief agency Île de Paix, came along to Taoudenni on my permit as well. Ali, Buoni, Moulay, Sidi Mohammed, and all the other competent guides were off on caravans, so we were stuck with old Hamd'r Rahman as our leader. He was nearly blind and had only three yellow stubs of teeth left, and he had a talent for getting his convoys stuck in the sand, finding the highest and steepest dunes to cross, and making every trip take twice as long as it should.

"Magnifico!" Paolo exclaimed, when I told him all this. "That way we'll see more of the desert."

I decided to take Boudj along in case I needed any help digging the cars out of the sand. Paolo and Viviane weren't used to hard labor, and Maurice had a broken hip that precluded him from doing any heavy lifting at all. We prepared a good supply of camel jerky for the trip, since it was light, compact, and practically imperishable.

The entire 500 kilometers north, the sand was almost as hard-packed as an asphalt road. There were no landmarks whatsoever, and since I didn't fully trust Hamd'r Rahman's abilities, I checked my compass constantly. I knew we were off course when we strayed far from the trail of camel dung left by caravans of years past, but it is an ironclad rule of the desert not to question the judgment of the guide.

We arrived at Taoudenni without major mishap. We had wandered longer than we would have under a more able pathfinder, and got stuck several times in sandpits, but even this had its positive side: In one such quagmire we collected a large pile of rocks to place under the wheels of the cars, and on closer examination found that these stones were actually petrified wood. Now that we knew to look for them, we found chunks of petrified wood for the next 100 kilometers. At one time, all this part of the desert must have been a living forest.

As we approached the mines, we passed many groups of men traveling with salt caravans in the other direction to their homes in the villages to the south. They were ragged and emaciated, mostly barefoot, and had a painfully long voyage over sharp rocks and sizzling sand ahead of them. Boudj handed a bowl of Araouane well water around to several of these miners, and told them they'd receive a warm welcome if they ever visited our town. The men gulped down the water gratefully. Up at the mines, they said, workers talk wishfully about the legendary wells of Araouane.

At Taoudenni, I finally got to see how the men of Araouane had been slaving away for generations. First we saw a row of little hills covered with what looked (at this distance) like busy ants. The workers crawled out of their holes as we pulled up, glad to have a diversion from the daily grind. An army patrol stopped us for questioning, but let us roam freely when I presented the letter from the governor.

The prison, shut down just two years earlier, had already been reclaimed by the desert. It was only a few roofless shacks built of salt blocks, not surrounded by walls, gates, or barbed wire. The desert itself was barrier enough.

A hundred or so miners and the few patrons supervising their labor came out to greet us. Many of them knew Boudj and me, since they often passed through Araouane on their way out here. They showed us all around the mines and let us crawl through the caverns hacked out of the ground. The miners explained that they dig square holes in the ground, about 30 feet on each side and 20 feet deep, until they come to the layers of salt. Then, with crude hatchets and adzes, they cut out slabs a foot wide and a yard long. The work is least onerous when the hole is fresh, but once they've removed all the salt at one stratum they start tunneling out in all directions. They dig and dig without any beams to support the tons of earth above, often, apparently, until the tunnel collapses.

I saw two of the Timbuktu laborers whom I'd hired to come to Araouane during my first year. Hammoudi, who spoke passable French, told me he'd learned a lot during his time in the town.

"After my work for you was over," he said, "I no longer got into debt. I'm now working at the mines on my own account, and get to keep the profits from the salt I cut." One of the others told me that Hammoudi secretly traded bars for carpets brought in by Algerian smugglers. He was, in his own way, an entrepreneur.

The men here lived in hovels built of salt too brittle to be transported. Their primary food was camel meat, from the caravan beasts who were too weak to make the return trip through the desert. The huts provided no protection from the region's harrowing extremes of temperature—in this part of the desert the difference between day and night can be 100 degrees Fahrenheit.

I'd been given advance warning not to drink the water at Taoudenni, but one look in the filthy drinking vats would have sufficed. The stuff was a rust-brown slime with the consistency of oil. Fortunately, we'd brought enough water from Araouane to last until we got back.

Paolo drove the miners crazy with thousands of questions, not one of which (I am sure) they even understood. He tried to tell them that, instead of mining salt, they should just go to the beach and collect it from the ocean. Boudj tried to translate, but since neither he nor any of the miners had ever seen the sea, the point was entirely lost.

Maurice and Viviane returned to Timbuktu, but Paolo came back to Araouane with me. By this time I'd grown quite fond of him. One day shortly after our return, we were eating lunch together at my house. It was so hot out that sweat was dripping from our foreheads onto our plates. All was quiet. Not even Paolo had anything to say. I looked at my watch, thinking that I wouldn't even wash the dishes in this heat, but would just leave them until evening and take a siesta. When I glanced at the dial, I realized that it was April Fool's Day.

"By the way, Paolo," I said, "this morning the kids told me that one of the famous Mauritanian caravans is near the village."

"What is so special about a Mauritanian caravan?"

"You know, the caravans from that small tribe that trains its camels to sit up on their haunches like German shepherds."

"What are you talking about?"

"You never heard of them? You've been in Africa such a long time, and never once heard of the famous begging camels?"

"No."

"In this region, everybody knows about this tribe but few people have actually seen them. They always keep their camels far away from any other herd, because they're afraid that if their animals see others lying down normally, they'll forget all their training."

"I've never heard such a thing."

"They came by last year, and I got some interesting pictures of them."

"Really?"

"A very odd tribe. They leave their camels out in the desert and carry water to them."

"Where are they now?"

"Salim told me that they were behind the second range of dunes off to the northwest of the village. Hussein helped them carry water this morning, and they gave him a goatskin of milk as payment."

"Do you think they are still there?"

"I can't imagine that they could have watered all the camels already, so I'm sure they are."

"I'll go and have a look."

An hour later, after unsuccessfully trying to nap, I was getting worried. Paolo wasn't back yet, and it was so hot that you could have roasted a piece of meat just by putting it down in the sand. I got up and started walking toward the dunes—at least, I tried to walk but spent most of my time hopping around instead. The sand burned my feet right through the toe holes of my sandals, so I went back and put on real shoes. Near the top of the first range of dunes I caught a glimpse of Paolo, trudging away in the distance.

"Paolo!" I cried out. "Paolo! Paolo!"

I kept on screaming until he noticed me.

He raised his arms, as if to ask me where the famous caravan was.

"*Primo Aprile!*" I yelled.

He jumped up and down. It definitely looked like he swore, but the sound didn't carry that far.

The kids were scandalized by my deception of Paolo.

"You lied!" they said. "You are always telling us that in Araouane nobody is allowed to lie, and now you go and do it yourself."

"Only because it is April first," I replied. "It's almost like a law, in many parts of the world, that on this one day you can lie or trick as much as you please. Just so long as you don't hurt or cheat anyone, that is. Today is April Fool's Day, so you *have* to make a fool out of somebody."

From that day on, the kids plotted and planned how to put one over on me next April first. But they weren't very skilled at deception. They were so pleased with their cleverness that they had to share their trick with me.

"Next April Fool's," they told me, "we will tell you there is a caravan in which the camels walk backward, and you will walk far into the desert to look for it."

"You kids aren't thinking," I said. "If you tell me the trick ahead of time, how do you expect me to fall for it?"

I went back to America, as I always did in mid-April. When I returned in October of the third year, bringing Emilie with me for the first time, the kids quickly enlisted her in helping to plot their big trick. The strategizing reached its climax just before March 25, my fifty-third birthday. They had organized a surprise party for me, and the buzzing excitement in camp would have alerted a corpse that something was up.

That morning, however, the aid organization Île de Paix radioed from Timbuktu with an urgent message. We were used to hearing from them several times a week, but today there was none of the usual chitchat.

"Araouane, Araouane, this is Timbuktu, can you hear me?"

"Affirmative, I get you five in five."

"You have to come to Timbuktu right away."

"Roger, we'll be coming down in about ten days."

"Negative, you have to come right away."

"What's the hurry?"

"I have orders from the highest authority that you have to come to Timbuktu at once. I can't tell you more—everyone in the desert can listen in on this transmission."

It sounded important. I knew the operator, a Frenchman named Henri Gall, and he was not a man to panic over nothing.

"Okay, then," I said, "we'll get the car ready and come down tomorrow."

"Negative, right away, right now."

Emilie was in the courtyard scraping the flesh off a gazelle skin. A few months earlier a nomad had brought her the little animal as a pet, but it had died yesterday and Emilie wanted to preserve its pelt.

"We have to go to Timbuktu right now," I said. "Henri can't tell us why, but we have to leave."

For months we'd been hearing, through shortwave BBC bulletins, of the growing unrest in the Sahara. In addition to the ever-present Tuareg uprising, there had been sporadic looting and burning in cities all over Mali. Just this morning the shortwave announcer had passed on a rumor that the Malian government had been toppled. If this was true, and if the various rebel groups wanted to take some Westerners hostage, we might be sitting ducks out here in Araouane.

So far none of these disturbances had actually touched Araouane. But now, it seemed, our luck had run out.

Getting out of Africa would be no piece of cake. We pondered ways of crossing to another country with the Land-Rover, but no route was feasible: We couldn't go east to Niger, since we had no visas and the only places to get the proper stamps were Bamako and Ouagadougou in Burkina Faso, which were both cut off by the antigovernment rioting; the Ivory Coast and Guinea were cut off by the same riot-torn areas. We couldn't travel west to Mauritania or north to Algeria, since the Gulf War was still on, and these nations were rumored to be turning away all American travelers. That left Timbuktu as our only exit point.

"Perhaps they've arranged an airlift of the Timbuktu foreigners," I said. "There are only fourteen, so it wouldn't be hard to charter a small plane. We'd better prepare to be gone a while."

In two hours we had packed. Bou-djema and Baba Boatna had changed the oil and filled the gas tank of the Land-Rover. I gave the assembled villagers an impromptu speech, and told them to continue their work in the garden no matter what happened.

"I promise you all that we'll never let you down," I said. "We will come back one day. In the meantime, you must demonstrate how strong Araouane now is. Like I've said, you can get by without any help from the outside world. Whatever problems come up, I know you will be able to solve them." A few of the women cried. Emilie cried with them.

Just before we left, Boudj whispered in Emilie's ear.

"Emilie," he said, "don't forget to tell Ernst what we would have done to him on his birthday and on April Fool's Day."

On the drive down to Timbuktu Emilie was practically in tears, for the sake of our friends in Araouane. She worried out loud about their fate, bemoaned our separation from them, but never once fretted about our own safety.

A few hours after we left the village, she told me what schemes they had concocted for the coming weeks. On my birthday, the children had arranged for Emilie and me to take two camels into the dunes by ourselves. We'd ride until sunset, and by "accident" wind up at a site where we would find a carpet laid with a fine dinner cooked by Babaya's household. Emilie had even managed to smuggle in a bottle of Châteauneuf-du-Pape from Bamako.

On April Fool's, Emilie would make a nice dinner, start to set a romantic mood, and then turn out the light and tell me to come to bed. I'd grope through the darkness and slip between the sheets, only to find Bou-djema there in Emilie's nightgown. Then all the other kids would burst into the room and shout, "*Primo Aprile!*"

When Emilie told me all this, I felt like crying too. I myself had come to Araouane as a visitor, but I left it as a member of the family.

HELL OR HIGH WATER

Emilie is sitting on an old camel whose neck is scarred with the marks of many previous owners. Her feet are covered with huge blisters, some of which have burst. Under the loose flaps of skin, sand and blood have mixed to form a dark paste.

I am being pulled behind the camel by a rope, like a child on a ski-slope T-bar. I can barely lift my bloody feet, so they drag over rocks and through patches of thorns. Both of our faces have been burned tomato-red by the sun.

The two of us and our cameleer, Sidi Ali, have been plodding along for two days and 110 kilometers on 15 liters of water and one pair of flip-flops, which we take turns wearing. Big flecks of smelly foam from the exhausted camel's mouth spray back at my face, but I am too tired to wipe them away. The beast's big penis with its wiggly little corkscrew tip hangs down from his belly like the hammer of a church bell. Since yesterday he has not had enough strength to retract it.

Finally we see Timbuktu in the distance. Emilie turns around slowly on the camel.

"Happy Thanksgiving," she says, smiling weakly. "And, if I've counted right, happy six-month wedding anniversary."

Everything has turned out so differently from what we had hoped and dreamed. The previous spring, after our radio summons, we had managed to get from Araouane to Timbuktu, and from there to Bamako in a car convoy of evacuating foreigners and onto a regularly scheduled flight out of the country. We had intended to go back to the States for the summer anyway, and we assumed the unrest would blow over by the time we planned to return.

Back in America, we'd sent out invitations for a magnificent wedding celebration to be held in Araouane the following winter. Officially we'd gotten married a year earlier at Manhattan's City Hall, but now we wanted all our friends around the world to join us in marking the occasion. We planned a gala week-long festival, complete with feasts of whole roast camels and vegetables from the garden, to culminate in a renewal of our wedding vows on Valentine's Day. More than sixty guests from Europe, Australia, and America promised to attend, and we even persuaded Swissair to run a special charter flight from New York to Timbuktu. The event would celebrate not only our marriage, but the successful revival of Araouane.

We spent the summer traveling in Australia, studying the natural vegetation of the outback and collecting seeds of the desert plants that had long provided the Aborigines with food. We figured that "bushtucker," the Australian term for food gathered in the wild, might flourish in the Sahara, and thus provide the population with a reliable source of food.

The trip was a great success. We found several knowledgeable Outback farmers who gave us all the help we needed. In no time we had gathered a supply of seeds and detailed instructions for planting them.

We ate kangaroos, crocodiles, lizards, giant maggots, the famous witchetty grubs, and even Australian camels. We spent time with Aborigines, learning how to harvest, prepare, and eat the bushtucker. We slept under the magnificent southern sky, with the Southern Cross above us. We swam and fished in the Pacific Ocean, the Indian Ocean, and the Arafura Sea. We were even tempted to settle on a 650,000-acre ranch which the owner offered to sell us for thirty cents an acre.

On the way home we spent a few days in Bali. It was a honeymoon of sorts, between our legal wedding in New York and the "real" one in Araouane.

Bad news was waiting for us when we got back to the States, in a letter from Mohammed Ali. In May, he wrote, barely a month after we'd left, the village had been attacked by armed Tuareg rebels who had taken everything of value from the project. The new Land-Rover, the generator, the medical supplies, the tools, the food, our cameras, video recorder, typewriters, compasses, binoculars, forks, spoons, and knives—all of it was gone. The only thing they'd missed was the radio transmitter.

The guerrillas had not hurt any of the villagers, and Mohammed Ali noted that everyone had worked to keep the garden from dying. At least the rebels had not been able to take *that* away. The people of Araouane had learned to take care of themselves, and they continued to struggle on in adversity.

If our friends in Araouane could be so brave, we knew we certainly couldn't give up.

My brother Peter found a new diesel Landcruiser for us. Emilie and I went on a shopping spree to replace all the things the rebels had stolen. We loaded up on too many delicious Swiss meals, and before we set off, I saw snow for the first time in three years.

This time, we experienced a new hostility toward foreigners as we passed through the Sahara. In Algeria children pelted our car with stones, and we had to replace one window with plywood. Africa showed us no welcome.

In the southern Algerian town of Tamanrasset, we visited the Malian consulate with a special introduction in hand: Back in New York, the country's ambassador to the U.N. had given us a letter instructing all Malian military and civilian authorities to assist us in getting the project back on its feet. The consul was friendly, but since the stretch of Sahara through which we'd have to travel was controlled by the Tuaregs, he had little power to help us. If we wanted any guarantee of safe passage, we'd have to contact the rebels themselves.

Through a series of intermediaries, we found our way to the secretary of the Mouvement Populaire pour la Libération de l'Azouâd. His name was Houssein Faradji, and he was introduced to us as a "refugee" from Mali. He was a scared, nervous man, who said that at any moment his life could easily end in a mysterious "accident."

"The Algerians will not openly accuse us of anything," he said. "After all, we are whites like them, and we are fighting the black Malians who are our traditional rivals. But still the Algerians do everything they can to prevent our victory. None of Mali's neighbors want to see us succeed, because they fear it would spark Tuareg uprisings in their own countries.

"The only person helping us," he continued, "is Muammar Qaddafi. We don't like to deal with such an unstable man, but what can we do? We have no other place to turn. Even the Islamic Alliance (and they certainly have enough money on hand) is against us, because some Tuaregs pretend to be Christian in order to get food from missionaries."

The plight of Houssein Faradji and his people touched us deeply, but we didn't understand why the Tuaregs would have any reason to attack *us*. His sad response was that all communication among the guerrillas had broken down. No Tuareg leader was in a position to issue a safe-conduct pass that would be honored by other bands, that many of the nomads would rob anyone who came through their territory.

"You haven't got a chance of getting from Tamanrasset directly to Timbuktu," he said glumly, "not one chance in a hundred. Someone along the way will take your car, your supplies, everything."

We left the sad man, very sad ourselves.

It seemed our only option was to take a 2,000-kilometer detour through Niger. This would involve crossing yet another country with a load of valuable cargo, and the country was one that under the best of circumstances had a bad reputation for official extortion. Niger, we'd heard, was a nightmare of heavily armed military checkpoints, at each of which soldiers stopped any passing car and demanded a hefty bribe.

In the stretch of desert near the border between Algeria and Niger, the binoculars were permanently fixed to Emilie's eyes. Lots of trucks had been attacked by the rebels in this area, so we were cautious to the point of paranoia. We hightailed it from any bluff or sand dune that looked even vaguely like it might conceal another vehicle.

It took a day and a half to get the necessary papers signed and stamped at the checkpoint. According to other travelers waiting for clearance, the rebels had recently robbed a dozen European tourists of their cars and belongings, stripped them to their underwear, and wished them bon voyage before abandoning them in the middle of the desert. Another party of French and Germans had formed a convoy that had been fired upon by the Tuaregs; during the night a Niger army patrol had come across the wounded men in their disabled vehicles and, thinking they were rebels, fired on them again. Three of the tourists were killed, all the others seriously wounded.

We passed the checkpoint and drove on. Just before nightfall several vehicles approached us from the south, blinking their headlights to signal us to stop. We had previously decided it was safer to ignore all such commands, to try to outrun any army or rebel patrols rather than face near-certain looting. When the vehicles came closer, however, we had to abandon any hope of flight: We saw that they were three armored personnel carriers, the first full of soldiers with grenade launchers, the second sporting a mounted cannon pointed directly at us, and the third containing a few dozen uniformed men staring at us over the barrels of their assault rifles.

We stopped, without even knowing just who they were. Since they were black, we knew they weren't Tuareg rebels, but we couldn't be certain they were legitimate army troops. One of the men approached our car and saluted. That was a good sign. He was a very small man with a friendly smile, which seemed oddly out of place. He politely asked if we were all right, or if we needed any assistance. Our relief was great, but so was our apprehension. We realized that if this encounter had taken place just a little bit later in the evening, we might not have seen their heavy armaments or answered their summons. We would have put the pedal to the metal, and probably been blown to smithereens.

Niger turned out to be an armed camp. We probably saw more automatic weapons than goats there. Our only casualty, though, was a bloody nose: At one checkpoint a soldier asked for my passport, and as I lowered my head to take it from my belt pouch, I banged into his gun barrel.

Generally, however, the authorities of Niger were quite friendly. Our letter of recommendation from the Malian ambassador proved very persuasive, even here. Since the government of Niger, like that of Mali, is black, its officers had no love for the Tuareg rebels, and their opinion of the guerrillas was the same.

"Kill them all," the officers would say, "and then the problem will be solved."

We entered Mali, crossed the Niger River by ferry at Gao, then drove down to Mopti to stock up on supplies. We bought nearly 12 metric tons of millet, sorghum, and baobab powder, and transported it by barge to Timbuktu.

The day we arrived in Timbuktu, the new governor demanded to see us. Lamine Diabirra, who had been such an enthusiastic supporter of our project, was in jail for supporting a failed countercoup. Despite the rampant food shortages brought about by the war and the government's chronic inability to feed its own population, the new governor showed little interest in our plans to help Araouane survive.

Through the social security office, he demanded $9,000 in back payment for "social security," but dropped the price to

$300 when I threatened to leave the country without donating another cent. Apparently he didn't like my project, but he was reluctant to give up a good target for present and future extortion.

The governor forbade us to travel to Araouane in the Land-Rover. The rebels, he said, would certainly take it from us. He permitted us to go by camel, with only a small load of supplies, so that when we got robbed, the guerrillas wouldn't profit very much. Before departing, Emilie and I had to sign a statement absolving the Malian government of all responsibility for our safety.

For several days Emilie repackaged our supplies into 50-kilogram bags for loading onto camels, and I went looking for a caravaneer. This turned out to be a very difficult task, since all the local nomads had long since fled into the desert. Timbuktu had turned into a ghost town. The luxury hotel Azalaï, where we used to stay while passing through the city, had no lights. The manager had run out of fuel for the generator, and had no money to buy more.

Panic and paranoia had set in. Anyone who had not fled was regarded with suspicion. Rumors flew out of control: Every rebel attack was reported in several versions, and modified or withdrawn as supposed victims surfaced. But the reality was equally horrific. Almost all the aid organizations had had at least one vehicle robbed, and if the drivers were members of the Bambara tribe, they were routinely killed.

The northern outskirts, where caravans had traditionally been arranged, were now totally deserted. The Moors with the herds of camels; merchants hawking grain, sugar, or tea; idlers gossiping about events on the trade routes—all were gone. Azou, one of the kids who used to roam the streets of Timbuktu, told us what had happened.

Last March the black population had gone on a rampage, looting and killing all the Tuaregs and Moors they could find. Their justification had been rumors of an alleged plot by the Arabs to launch a massacre of their own.

"I felt sorry for some of the Arab kids who were killed," he said, "because they had been my friends. But the looting was lots of fun. We didn't really get much out of it, since everything was destroyed, but it was fun anyway.

"I have never been swimming in water," he continued, mimicking the crawl, "but that day I was swimming in people. There was torn fabric all over the street, and sand mixed with rice, millet, and cornmeal. The army had told us the Arabs were armed, so we went through their houses and broke everything. We made some of the Moors and Tuaregs dance before we killed them."

A few nomads hung cautiously around the outskirts, ducking into the desert whenever gangs of blacks came near. The government had organized groups of vigilante *"moniteurs"* who were theoretically supposed to report any suspicious activity, but sometimes lynched any Moors or Tuaregs they found, either clubbing them to death or making them dance before shooting them.

With the help of a very dark-skinned Arab ex-Araouanite named Malik, I persuaded one of the few remaining Moors to talk with me about hiring camels, but the negotiations were brief and futile. He demanded double the normal price, and that for a caravan going only halfway to Araouane. I didn't want to wind up looking for a new ride in the middle of the desert, so I declined. The next nomad I found, a wrinkled old man in a tattered *boubou,* insisted that I leave most of my supplies behind, since he didn't have enough camels for them. All of my old friends among the nomads were gone, and it was clear that none of these strangers wanted to have anything to do with me.

At last I managed to locate a man willing to make the trip, and at the usual rate of payment. Sidi Ali Ould Raïs, a small, skinny Moor in a dark blue *boubou,* understood not one word of my classical Arabic, but his intermediary spoke fluent French. Everything was readily agreed to: the number of camels, the weight and volume of baggage, the price, the time and place of departure.

When the time came to leave, we unveiled a special contribution to the caravan: a camel saddle we'd brought with us from Alice Springs, Australia, a metal frame fitting neatly on either side of the hump, providing a comfortable, secure seat in front and a rack for luggage in the back. Everyone was amazed by the contraption. One old man stared openmouthed for several minutes, then took a step back in awe and fell right on his butt. He stayed on the ground, muttering that God would surely punish anyone who so irreverently altered what He had ordained.

Emilie had traveled from Araouane to Timbuktu by camel just nine months before, but I hadn't made the trip for four years. We both looked forward to a pleasant week of hiking in the desert, breaking in our new saddle, sipping tea by the campfire in the cool evenings, staring up at the star-studded sky.

We left Timbuktu the Sunday before Thanksgiving with Sidi Ali, his little half-brother Hamma, and ten camels. The boy had shiny, straight black hair and skinny legs that seemed to be made entirely of muscle and sinew.

This year the rains had again been excellent. The desert, which only four years earlier had been mere sand in most places, was now covered with all kinds of foliage. Knee-high *cram-cram* bushes, with their delicate white star-shaped flowers that looked pretty but made you itch horribly whenever you brushed against them, were everywhere.

But even out here, in the pure emptiness of the desert, the war was not far distant. Most of the wells had been destroyed, or were too dangerous to approach. When Hamma came back from a watering hole one morning, his sack contained only a pitifully small amount of jellylike liquid. Our midday tea had the appearance and consistency of crude oil.

The next morning we saw two Moors on camels. They were trying to make their camels gallop, and when they reached the crest of a dune, they hid behind a large bush.

"Even the nomads are spooked now," I said, laughing. "They think we're guerrillas."

The next minute I stopped laughing. Three men came charging from behind a dune ahead of us, two of them carrying machine guns and one shooting a pistol wildly in every direction.

The man with the pistol gesticulated for us to put our hands over our heads. Neither Emilie nor I had had a lot of experience with that kind of situation, but we quickly figured out that the first thing to do is obey any order given by a man with a gun.

The guy seemed crazy, or perhaps drugged. He jumped about wildly, shooting at nobody and nothing in particular. He wore some sort of ragtag uniform and, unlike the other two, had not covered his face with a turban in usual Tuareg fashion.

He stuck the pistol in his pocket and frisked me thoroughly. I noticed that he'd neglected to flick on the safety catch, and prayed he'd shoot his balls off.

When he moved on to Emilie, I felt a knot form in my gut. I knew that if he decided to mistreat her, I would be powerless to stop him. The other two kept their machine guns trained on me, just in case I didn't appreciate this fact.

As he approached, Emilie smiled and spoke to him in French.

"How do you do, Monsieur?" she said.

"I am fine, and how are you Madame?"

"Thank you, I am fine."

He frisked her gently and discreetly, without shooting off his pistol even once. Then he came back to me and told me to remove my suspenders, belt, pants, and shoes. Through it all I never even got a straight look at his face; my eyes were glued to the barrel of his .45 caliber gun the whole time. The black hole of the muzzle looked awfully big. I'm sure I wouldn't be able to recognize him if I saw him today.

The bandits tried to figure out how to use the suspenders— it's difficult to attach them to a *boubou*. Then they inspected the belt, looked curiously at the zipper in the back, and threw it on the pile of other loot. Had they played with the zipper a bit more they would have discovered $2,500 in the secret pocket.

As I tried to reason with our captors, I heard Emilie call to me in English. "Ernst, lower your arms a bit," she said. "The other money pouch is showing from under your shirt!"

I lowered my arms, but one of the men with a machine gun shouted at me and made as if to shoot me if I didn't get them back up. I contorted myself like a pretzel trying to keep my arms up and my shirt down. Emilie's warning saved us another $3,000.

And now she was patiently talking to our captors: "*Messieurs,*" she said, "*passeport, s'il vous plaît.*"

Amazingly, they gave it back to her. After a bit more pleading, they also returned my passport, my trousers, and my money belt. But there the amiability ended.

They took all of our camels but one, and started to move off. The pistoleer fired a few shots in the air and threatened to kill us if we followed them. They took Hamma, and as soon as they disappeared over the dunes, we heard two volleys of automatic fire and two pistol shots.

Sidi Ali stood like a statue.

We did not know whether they had actually killed Hamma or whether they merely wanted to frighten us. Emilie suggested that we wait and go back to check on Hamma, but that would have been suicidal: Either he was already dead, or the weapons fire was meant to warn us not to follow.

That is how we happened to be in the middle of the desert with one old camel, 15 liters of water, and one pair of flip-flops, with over 100 kilometers between us and Timbuktu.

I was really proud of Emilie. Terrorism and banditry had hardly been a part of her suburban upbringing, but she'd come through like a trooper. Not that I'd had much more experience with gunplay myself; apart from some youthful target shooting and a few minor encounters with old hunting rifles, I'd only looked directly down the barrel of a loaded gun once, long ago in Beirut, when a Palestinian hothead took offense at my liaison with an Arab woman.

We commiserated as we walked, hashing over each detail of our hijacking, although deliberately avoiding any mention of Hamma's fate to spare Sidi Ali further pain.

"All those prize seeds gone!" I lamented. "Our whole trip to Australia for nothing!"

"Our special saddle!" moaned Emilie.

"Our tools!"

"Our Thanksgiving dinner! The bottle of Bordeaux!"

"All our clothes, books, food . . ."

"Kurt's camel tenderizer," Emilie said, beginning to smile. "Oh, my God!" We looked at each other and, despite ourselves, began to laugh.

We'd told my brother Kurt, the food researcher, that most of the camels we ate in Araouane were old and scrawny. "No problem," Kurt said, "we have a meat tenderizer that can turn shoe leather into filet mignon."

A few days later we received a container of brown powder. The accompanying note explained that the vial contained only the tenderizer's active ingredient, that all the salt and spices had been removed to make it more compact for transport. "Mix one part of this tenderizer with one hundred parts salt and seasonings of your choice," the note read. "Just a tiny pinch is all you need, otherwise it will dissolve your guts."

Now, as we hobbled toward Timbuktu, Emilie and I imagined the nomadic thieves eating their victory meal with our fancy foreign condiment.

"It'll give them bigger holes in their bellies than any Kalashnikov could make," I said happily.

"*Bon appetit!*" said Emilie.

When we arrived in Timbuktu, we collapsed into the arms of Maurice and Viviane at the Île de Paix office. They had not slept since the previous day.

"We were so worried," Maurice told us. "The day before

yesterday we heard on the radio from Araouane that the rebels had attacked the village again, looking for you. When they found out that you weren't there, they got mad. They mistreated Mohammed Ali, even spilled his morning tea, stole his teapot, glasses, and everything else that could be moved. And we knew you were headed straight in their direction. Thank God you're here."

Sleep wouldn't come that night to us either. All our dreams of returning to Araouane had been quashed. Over dinner the local head of the Red Cross had presented us with a plan for Araouane's evacuation.

Emilie and I had talked about evacuation, but it didn't make much sense. Evacuation to where? To Timbuktu? We couldn't bring the Moors to town, since it would be open season on them. To Mauritania? There it would be open season on the blacks; the Arabs of Mauritania didn't even need a hunting license to shoot blacks, rumor had it.

There was no place that would be safe for Araouanites of both races, and we couldn't stand the idea of splitting the population by color. Babaya's black wife would have had to go to one camp, his white one to another. Everything the villagers had worked for together—the school, the garden, the trees—would be cast away. The Araouanites would end up just another bunch of starving refugees in a string of squalid shantytowns.

We spoke with the villagers by radio—miraculously, they still had the transmitter—and asked them to figure out what they wanted to do. Boudj gave us daily status reports, in perfectly fluent French. The Araouanites, he and Mohammed Ali told us, wanted to stay.

I was inspired by their spirit and courage. Through all the adversity, all the threats, all the disappointments, they had managed to keep going. The rebels had taken their goods, but not their determination.

The very fact that the guerrillas would choose to attack Araouane at all was proof of our project's success. The Tuaregs would never have considered raiding four years earlier, because there wouldn't have been anything there for them to take. Now, after three years of hard work, there was plenty of fresh food, as well as sturdy homes and thriving gardens to tear down. But they would never be able to destroy the villagers' greatest accomplishment—the knowledge that they could control their own destiny.

At the home of Maurice and Viviane we listened to shortwave radio reports from all over Africa. The whole continent seemed to be in turmoil. Sudan and Ethiopia were still killing fields. Famine wracked these countries, and Somalia as well. In South Africa, blacks were killing blacks. Riots and repression ravaged Kenya, Mozambique, Algeria, Madagascar, Zaire, and Nigeria.

On the eleventh of December, just after we'd finished dinner, we heard machine-gun fire in the distance and answering fire from right near our house. A grenade exploded immediately outside our window, rattling the wooden shutters. In the morning we found out that the rebels had attacked the governor's residence. Military discipline had completely broken down, and anarchy reigned. Men with automatic rifles roamed the streets, threatening to kill Red Cross officials and anybody else they bumped into.

We decided to haul ass. We left money for Araouane with Dramane Alpha, a Red Cross driver whom we'd known for years. All the foreign aid workers evacuated Timbuktu, except for a skeleton crew from the Red Cross.

The last direct communication we had with Araouane was a radio message the day of our departure. It was Boudj, wishing us a pleasant journey:

"Everybody is well in the village," the boy said. "Everybody is working, rebuilding the garden as you taught us. We wish you a safe trip, and hope to see you soon."

E p i l o g u e

November, 1992

It was nearly a year before I managed to return to
Araouane. We'd had no news from the village for months. For
a while, we'd had regular reports from Alpha (in exchange
for tubes of hemorrhoid ointment by return post). Mohammed Ali was working for the Red Cross in Timbuktu, and
occasionally managed to send food shipments up to the village with the nomads. But on a third raid, the Tuaregs had finally made off with the transmitter, and neither of them had
had any direct news from Araouane since, although they'd
heard a rumor that the solar pumps had ceased functioning.

I arrived in Timbuktu at the end of October, and shortly
before Thanksgiving, Mohammed Ali and I persuaded the authorities to let us go up by Land-Rover with two heavily
armed rebels from the cease-fire commission as escorts.

When we first glimpsed the village in the distance, I could
barely contain my excitement. We entered from the east, by
the mosque, then pulled up to Babaya's house, to show respect. A mob of kids swarmed around the car, but the only
one I recognized was little Hussein. "Where are Boudj, Salim,
and all the others?" I asked.

He mumbled something unintelligible, and looked
ashamed.

"Come on," I said, assuming they merely hadn't heard the
engine's noise. "Where are they?"

"In Taoudenni," Hussein said. "All of them."

Babaya and a few old men approached. After the usual formalities, he confirmed Hussein's report.

"Everyone who could work had to go to Taoudenni," he
admitted. "They had to cut salt, or else we would all starve."

"But what happened to all the food Mohammed Ali sent
up?" I cried. "I bought plenty of camels, and enough supplies
to keep the whole village fed!"

Blank stares from the old men, and a shrug from Babaya. Since our last meeting, his ample belly had gotten noticeably bigger.

Old Tata rushed up and chattered, in her usual noisy way, "Welcome, Aebi, welcome. Why have you been away for so long?"

I spotted old Baba Cambouse. "What is going on here?" I asked him. "What happened?"

"Lots of problems, lots of problems," he said. He turned away, unable to look at me.

I walked through the project, taking stock of the damage. Sand had encroached on all the windbreak walls. The grape vines and virtually all the trees were dead and withered. Only a few of the olive and fig trees still showed a little green around the trunks. They were alive, but barely.

My house had been looted thoroughly. The carpets, mats, dishes, cameras, and clothes had been stolen, the shutters and walls riddled with bullet holes. All the furniture had been hacked to pieces, and the control panel for the solar collector had been smashed to bits. The rebels showed their contempt in what they decided not to steal: stacks of shredded books, maps, and papers, a mess of dried vegetables, and two bottles of wine that Emilie had smuggled in for my aborted birthday party.

In Mohammed Ali's schoolhouse, a pile of ashes still lay where the Tuaregs had burned textbooks to boil their water for tea. Baba Boatna followed me from place to place, but when I spoke to him in French, the boy no longer seemed to understand.

We were joined by Araouata, who admitted that my food shipments had indeed reached the village, but explained that Babaya had refused to distribute them. Many of the Arabs' extended families had taken refuge in the town, and they were the ones who got the supplies intended for all who worked in the garden. Without any rations to sustain them, the black villagers stopped tending the crops, and had no choice but to lapse into slavery again. Araouata said he'd tried to reason with the Moors and was ashamed of his failure.

When night fell, the rebels who'd escorted us from Timbuktu insisted that we lock the Land-Rover. It was the first time I'd ever locked a car in Araouane.

The next morning, Baba assembled the remaining inhabitants in the theater. All the old blacks were there, and most of the Arabs. The blacks sat behind the Arabs. Only old Baba sat next to me.

There wasn't much to say. I told them I had neither the desire nor the stamina to start all over when the trees we'd already planted would probably have started giving them the fruits of their labors—if only they'd been kept alive.

The Arabs nodded gravely, the few elderly blacks diverted their eyes, and the children giggled. Salah Sultan, the village chief who had finally moved up from Timbuktu to escape anti-Arab bloodshed, covered his face with his turban and went to sleep.

I said that there was a chance that we would resume the project when the political situation allowed, but they'd have to earn that chance. They would have to make every attempt to save whatever trees were not yet dead and to protect the garden from the accumulating sand, just as they did with their own houses.

Old Baba told me that he had not slept all night. "I am very, very ashamed, but there is not much I could have done, I am only an old man," he said. "By the way," he added, "this year again we had not a drop of rain. Maybe it's because we didn't keep up the garden." He shuffled off.

I wanted to go. Everything in the village was covered with clouds of flies, everything looked ugly. I could hardly look at anything or anyone. I felt a huge void in me, a painful emptiness, a big blank white space that wasn't on any map.

In the evening, Salah Sultan invited us all for dinner. Not to make matters worse, we went. Hababou, Amma's father, served us, no black man sat with us, nobody talked, and the flies buzzed ceaselessly. I'd even lost my taste for Araouane food, even though Hababou had remembered my favorite dishes and had prepared them well. After the third glass of tea, we went to sleep. I heard one of our escorts swear. "This fucking village doesn't even have any cigarettes."

But I could not help dreaming that the trees would grow again in Araouane. We had planted seeds in lifeless sand, and they had grown. There was such good, rich soil in the hearts of so many of the villagers, I couldn't help but hope.

I had an idea. Because of the unrest, the nomads no longer dared venture into the towns to sell their camels, goats, and sheep. As a result, the herds had gotten huge—I'd seen them on our journey, sometimes filling the whole horizon. Such an abundance of food, and yet the country was dependent on foreign food!

Back in Timbuktu, I gave Mohammed Ali $2,000 in West African currency. To Alpha I gave $600, the generator, the toolboxes, and the clothing Emilie and I had left in Timbuktu on our disastrous previous trip, to sell for additional cash. Buy livestock from the nomads, I told the pair. Slaughter the animals, and dry the meat in the sun. Sell it for local relief instead of relying on sardines from Japan and corned beef from Finland and Denmark that got waylaid by the wealthy anyway.

They liked the idea. Mohammed Ali, being an Arab, could venture out into the desert without fear; Alpha, being black, could take the meat to market in Timbuktu. If all went well, they'd make Araouane the gathering point for trade with the nomads. I hope you get filthy rich, I told them, because that way you'll know it's working.

In the meantime, Mohammed Ali promised to keep me posted on how the village fares. There wasn't much to work with—just a few stalks of half-withered olives and pomegranates—but then there hadn't been much to begin with. And maybe, just maybe—in a year or two, when the Kalashnikovs had gotten jammed with sand and life had settled down—the seed of independence would take root again. If I get the green light, I'll be back.